THERE IS LIFE

AFTER THE

NOBEL PRIZE

Eric Kandel receiving the Nobel Prize in Physiology or Medicine,
Stockholm, Sweden, December 10, 2000
©The Nobel Foundation, photo: Hans Mehlin

ERIC R. KANDEL

THERE IS LIFE AFTER THE NOBEL PRIZE

Columbia University Press

New York

Columbia University Press
Publishers Since 1893
New York   Chichester, West Sussex
cup.columbia.edu

Library of Congress Cataloging-in-Publication Data
Names: Kandel, Eric R., author.
Title: There is life after the Nobel Prize / Eric R. Kandel.
Description: New York : Columbia University Press, [2022] | Includes
bibliographical references and index.
Identifiers: LCCN 2021021959 (print) | LCCN 2021021960 (ebook) |
ISBN 9780231200141 (hardback) | ISBN 9780231553469 (ebook)
Subjects: LCSH: Kandel, Eric R. | Columbia University—Faculty—Biography. |
Howard Hughes Medical Institute. | Neuroscientists—United States—
Biography. | Medical scientists—United States—Biography. | Nobel Prize
winners—United States—Biography. | Neurosciences—Research. |
Molecular neurobiology—Research. | Memory—Physiological aspects—Research.
Classification: LCC RC339.52.K362 A33 2022 (print) | LCC RC339.52.K362
(ebook) | DDC 616.80092 [B]—dc23
LC record available at https://lccn.loc.gov/2021021959
LC ebook record available at https://lccn.loc.gov/2021021960

Columbia University Press books are printed on permanent and durable
acid-free paper.
Printed in the United States of America

Cover design: Milenda Nan Ok Lee
Cover image: Eve Vagg

To Denise,

my wonderful wife and my constant source of inspiration

CONTENTS

INTRODUCTION 1

1 MOVING TO COLUMBIA AND THE HOWARD HUGHES
MEDICAL INSTITUTE 5

A Biological Basis for Psychiatry 7

2 FURTHER ADVANCES IN SCIENCE 13

How Long-Term Memory Is Stored in the Brain 14

Uncovering the Biological Basis of the Gateway Effect
in Drug Addiction 19

Identifying Dopamine's Role in the Cognitive Symptoms
of Schizophrenia 23

Ameliorating Age-Related Memory Loss 25

3 ADVENTURES IN THE PUBLIC UNDERSTANDING
OF SCIENCE 29

In Search of Memory:
The Emergence of a New Science of Mind 31

CONTENTS

The Brain Series 35

1. An Overview of the Brain 36
2. The Patient Speaks 40
3. Brain Science and Society 42

*The Disordered Mind: What Unusual Brains
Tell Us About Ourselves* 53

4 INTRODUCING BRAIN SCIENCE TO ART 57

*The Age of Insight: The Quest to Understand
the Unconscious in Art, Mind, and Brain,
from Vienna 1900 to the Present* 58

*Reductionism in Art and Brain Science:
Bridging the Two Cultures* 63

5 RETURN TO AUSTRIA 71

6 COLUMBIA UNIVERSITY AND THE SCIENCE OF MIND, BRAIN, BEHAVIOR 75

The Kavli Institute for Brain Science 76

The Mortimer B. Zuckerman Mind Brain Behavior Institute 77

The Jerome L. Greene Science Center 79

CONCLUSION 83

CONTENTS

Acknowledgments 85

Appendix: Awards 87

Notes 97

References 101

Index 105

INTRODUCTION

Since the early 1960s my wife, Denise, and I have spent much of each August in the town of Wellfleet on Cape Cod. In 1976 we bought a summer home in the area; the house has a beautiful view of, and direct access to, Wellfleet Bay and is a lovely place for boating and swimming. I also enjoy playing tennis on the superb clay courts at Oliver's Tennis Club, a ten-minute drive from our house. Most important, August is an opportunity for Denise and me to have a peaceful vacation and to get together with our two children, Paul and Minouche, and their families, who come for at least part of the month.

One day in August 1996, as Denise and I were outside hanging the laundry up to dry, the phone rang inside the house. I went to answer it and found Stephen Koslow, my program officer from the National Institute of Mental Health, on the line. He told me that the grant I'd applied for had been awarded and should be funded within a reasonable period of time. He then added that he and a number of people at NIMH thought I would get the Nobel Prize.

FIGURE INT.1 Denise and me on our wedding day, June 10, 1956.

After I hung up the phone, I went back outside and continued to hang the laundry. I told Denise that Koslow thought I was going to get the Nobel Prize. To my astonishment she responded, "I hope not soon."

I turned to her and said, "What do you mean 'not soon'? How could you, my wife, say that?"

Denise reminded me that when she was a graduate student in sociology at Columbia, she worked with Robert Merton and Harriet Zuckerman, who studied people who had won various prizes, including the Nobel Prize. They found that, by and large, after someone has won the Nobel Prize, he or she does not contribute much more to science. They become too occupied with ceremonial and social activities. "You still have a few ideas," she said. "Play them out. There is lots of time for the Nobel Prize later on."

So four years later, on October 9, 2000, when I received a telephone call from the Nobel committee telling me that I had been awarded the Nobel Prize in Physiology or Medicine for my work on the biological basis of learning and memory, I realized that in addition to the pleasures and gratification of this remarkable award, I now had a challenge on my hands: I had to prove to Denise that I was not yet completely dead intellectually.

My aim in writing this book is to affirm that winning the Nobel Prize does not presage one's intellectual demise. Quite the contrary! It can spark new and unexpected creative endeavors and experiences that coexist in harmony with continued progress in one's scientific work. In the following pages I describe the post-Nobel years of my life—my very enjoyable and, I like to think, productive research years in neuroscience at Columbia University, as well as the opportunity I had to help shape the future of interdisciplinary research at Columbia's Mind Brain Behavior Institute. I also describe the wonderful new experience of imparting knowledge about mind, brain, and behavior to the general public, a public that, I now appreciate, is eager to learn and to know more about brain science and brain disorders.

CHAPTER 1

MOVING TO COLUMBIA AND THE HOWARD HUGHES MEDICAL INSTITUTE

I n 1974 I was invited to move from New York University, where Alden Spencer, James Schwartz, and I had formed the nucleus of the Division of Neurobiology and Behavior, to the Columbia University College of Physicians and Surgeons, where I would be founding director of the Center for Neurobiology and Behavior.

The move was attractive to me for several reasons. Historically, Columbia had a strong tradition in neurology and psychiatry, and a friend and former clinical teacher of mine, Lewis (Bud) Rowland, was about to assume the chairmanship of the Department of Neurology. In addition, my first experience in neurobiology had been at Columbia. In 1955, while finishing medical school at NYU, I decided to take an elective course in basic neuroscience at Columbia with Harry Grundfest. At that time, no one on the faculty at NYU was doing basic neural science, and Grundfest was the most intellectually interesting neurobiologist in New York. The experience of working with him altered my career and my life. Several years later, when Grundfest retired, I was recruited to replace him.

Finally, Denise was on the Columbia faculty and our house in Riverdale was near the university, so a move to Columbia would greatly simplify our lives. I decided to leave NYU and was able to persuade Schwartz, Spencer, and Irving Kupfermann to join me. In 1983 I became a University Professor at Columbia.

In 1984 Donald Fredrickson, the newly appointed president of the Howard Hughes Medical Institute, asked Schwartz, Richard Axel, and me to form the nucleus of a Howard Hughes Medical Institute at Columbia devoted to molecular neural science. The institute gave us the opportunity to recruit from Harvard both Thomas Jessell and Gary Struhl, as well as to keep Steven Siegelbaum at Columbia. So I resigned as director of the Center for Neurobiology and Behavior to become a Senior Investigator at the newly formed Howard Hughes Medical Institute (HHMI) at Columbia University.

HHMI came into its own as a research institute after the death of business magnate Howard Hughes and the sale of the Hughes Aircraft Company, which was owned by the institute and provided its remarkable endowment of $5.2 billion. That endowment was more than $1 billion larger than the Ford Foundation's, which until then had been the world's largest medical research endowment. Frederickson and the institute's trustees set out to fund, in a unique way, four areas of biomedical research, one of which was neuroscience. The institute did not hire scientists to work at a campus of its own, nor did it award grants. Instead, as it did at Columbia, it recruited outstanding scientists and funded their work in their own labs. Thus, HHMI not only paid our salaries and those of the people employed in our lab, it also covered overhead costs. As James B. Wyngaarden, a former director of the National Institutes of Health (NIH), said: "There are a number of things Hughes can do that we just can't do. With their new assets, they are in a position to be a very powerful force."

The institute had a twofold objective in seeking a better understanding of how the brain works: first, to prevent, ameliorate, or cure brain disorders; and second, to provide a clearer physical basis for the practice of psychiatry. HHMI realized that to achieve these objectives, scientists would need longer-term funding than they had been accustomed to receiving. Extended, renewable funding would encourage them to work on ambitious problems that would take more than a year or two to produce results. I felt this myself. With funding from NIH, I would have worried about how likely I would be to get support if I didn't get results in one or two years. But HHMI's longer view let me take more of a chance, rather than doing predictable things just to get results.

A BIOLOGICAL BASIS FOR PSYCHIATRY

Understandably, the first objective in improving our understanding of the brain was to prevent, ameliorate, or cure brain disorders. The importance of the second objective—to provide a biological basis for the practice of psychiatry—may appear less obvious. However, that objective has been a concern of mine ever since I left my postdoctoral training in neuroscience at NIH and started my residency in psychiatry in the early 1960s.

When psychoanalysis emerged from Vienna early in the twentieth century, it represented a revolutionary way of thinking about the human mind and its disorders. Under the influence of psychoanalysis, psychiatry was transformed in the decades following World War II from an experimental medical discipline that was closely related to neurology into a nonempirical specialty focused on psychotherapy. In the 1950s academic psychiatry abandoned some of its roots in biology and in experimental medicine and gradually became a therapeutic discipline based on psychoanalytic theory. This drift away from biology was not due simply to the changes in psychiatry; it was also due in part to the slow maturation of the brain sciences. The changes in academic psychiatry, of course, greatly influenced the training of psychiatrists and the practice of psychiatry.

By 1998 major developments had taken place in the brain sciences, particularly in our understanding of how different aspects of mental functioning are represented by different regions of the brain. These advances presented psychiatry and cognitive psychology with a unique opportunity to help biologists in the search for a deeper understanding of the biological basis of behavior. In support of this possibility I wrote a paper entitled, "A New Intellectual Framework for Psychiatry."[1] This was an outline of how psychiatric thinking and the training of future practitioners might be aligned with modern biology.

The outline emphasized two things. First, future psychiatrists must have a greater knowledge of the structure and functioning of the brain than was available in most training programs at that time. And second, analysis of the interaction between social and biological determinants of behavior—the purview of academic psychiatry—can best be carried out by practitioners who understand the biological components of behavior.

I identified five principles that, put very simply here, encompass the way biologists think about the relationship of mind to brain.

First, all mental processes, even the most complex psychological processes, derive from operations of the brain. As a corollary, behavioral disorders that characterize psychiatric illness are disturbances of brain function, even when the causes of the disturbances are environmental.

Second, genes and their protein products play an important role in determining the structure and function of the circuits that neurons form in the brain; it is in this way that genes and combinations of genes exert significant control over behavior. As a corollary, genes contribute to the development of major mental illnesses.

Third, altered genes do not, by themselves, explain all of the variance of a given major mental illness. Genetic, social, and developmental factors operate together in a feedback loop: combinations of genes contribute to behavior, including social behavior; and behavioral and social factors modify the expression of genes and thus the function of nerve cells. In other words, learning produces alterations in gene expression.

Fourth, alterations in gene expression that result from learning give rise to changes in neural circuits. These changes contribute to the biological basis of individuality and may be responsible for socially induced abnormalities of behavior.

Fifth, to the extent that psychotherapy or counseling is effective and produces long-term changes in behavior, it presumably does so through learning; that is, by producing changes in gene expression that alter the strength of synaptic connections and structural changes that alter neural circuits in the brain.

Fortunately, some people in the psychoanalytic community thought that empirical research was essential to the future of the discipline. Because of them, two trends have taken off in the last several decades. One is the insistence on evidence-based psychotherapy; the other is an effort to align psychoanalysis with the findings emerging from studies of the biology of mind.

Perhaps the most important driving force for evidence-based therapy has been Aaron Beck, a psychoanalyst at the University of Pennsylvania. His work with depressed patients led him to develop cognitive behavioral therapy, a systematic approach to therapy that has enabled him and others to study empirically the outcomes of treatments for depression. His studies showed that cognitive behavioral therapy is as effective as, or more effective than, antidepressant medication in treating people with mild and moderate depression. It is less effective in severe depression, but it acts synergistically with antidepressant medication.

One reason we know so little about the biology of mental illness is that until recently we knew relatively little about the neural circuits that are disturbed in psychiatric disorders. That said, we have now identified a complex neural circuit that becomes disordered in people with depressive illnesses. One part of this neural circuit is the right anterior insula. In a study of people with depression headed by Helen Mayberg of Emory University, each person received either cognitive behavioral therapy or an antidepressant medication. People who started with less than average activity in the right anterior insula responded well to cognitive behavioral therapy but not to the antidepressant. People with greater than average baseline activity responded to the antidepressant but not to cognitive behavioral therapy. Thus, it was actually possible to predict a depressed person's response to specific treatments from the baseline activity in their right anterior insula.

These results show us several important things about mental disorders. First, the neural circuits that are disturbed are likely to be complex. Second, we can identify specific, measurable biological markers of a mental disorder, and those biomarkers can predict the outcome of two different treatments: psychotherapy and medication. Third, psychotherapy is a biological treatment, a brain therapy. It produces physical changes that can be detected with brain imaging.

CHAPTER 2

FURTHER ADVANCES IN SCIENCE

A fter being awarded the Nobel Prize in 2000 for the discovery of how short- and long-term memory are initially *formed* in the brain, I decided that my next step would be to explore how memory is ultimately *stored* in the brain. Put simply, I wanted to find out how we remember what we have learned. After all, we are who we are because of what we have learned—our experience—but more importantly, because of what we remember.

So I returned to my clinical roots and expanded my work to include disorders of memory. Why disorders? Because studying disorders can give us useful insights into how the brain normally functions. In the course of this work, my colleagues and I uncovered a remarkable new aspect of the function of protein synthesis.

Memory is compromised in one way or another in many complex brain disorders. I decided to focus on the nature of memory loss in three of them: drug abuse, schizophrenia, and age-related memory loss. My colleagues and I created three types of genetically modified mice so that we would have an animal model of each of the brain disorders. We could then study changes in memory in each model not only at the behavioral level but also on the cellular and molecular levels. Studying complex disorders on such basic levels can reveal how various processes in the brain interact to create those disorders.

HOW LONG-TERM MEMORY IS STORED IN THE BRAIN

In our earlier studies of the giant sea snail *Aplysia*, my colleagues and I had found that the brain has separate processes for initiating memory storage and for maintaining that memory over time. Initiating memory does not require the synthesis of new protein, whereas maintaining memory does. Once we understood the basic mechanisms involved in the initial steps of forming a memory, we were in a unique position to ask how the brain stores memory for the long term. Kausik Si, who joined my lab in 1999, addressed this question in a remarkably original and important set of experiments.

Memory is initiated when a neuron, or nerve cell, responds to a stimulus by firing an electrical signal. The signal travels down the activated neuron's axon to its terminals. These terminals form contacts, called synapses, with other neurons, known as target neurons. The terminals of an activated neuron release a chemical substance, a neurotransmitter, that crosses the synaptic cleft— the tiny gap between neurons—and binds to receptors on postsynaptic target neurons, activating them in turn. This action—in which an electrical signal stimulates a presynaptic neuron to release a chemical neurotransmitter that then stimulates receptors in the postsynaptic neuron to produce an electrical signal, known as a synaptic potential—changes one or both of the cells and strengthens (or weakens) the synaptic connection between them. The modification of synaptic connections between neurons is the fundamental mechanism of memory formation.

Our earlier research in *Aplysia* had shown that the transition from short-term or intermediate-term memory to long-term memory storage requires the synthesis of new proteins (fig. 2.1). Moreover, long-term memory occurs only at synapses that have been activated, or "marked." Marking enables a synapse to capture and use gene products that are made in the cell body and shipped to all of a neuron's terminals. One such gene product is mRNA (messenger RNA), a copy, or transcript, of the DNA in the cell's nucleus that is used as a template for protein synthesis. The process of transcribing a DNA sequence into mRNA and the subsequent translation of that sequence into a protein is known as gene expression.

When Kelsey Martin worked in my laboratory,[1] she and her colleagues found two distinct components of synaptic marking in *Aplysia*: one that *initiates* memory storage through the growth of new synapses and another that *stabilizes* these functional and structural changes at the synapse for long-term memory storage. The stabilizing component requires, in addition to protein synthesis in the cell body, protein synthesis at the synapse.

Si argued that if a neuron sends mRNA transcripts to all of its terminals but only those terminals that are marked can use the transcripts to synthesize new proteins, then the shipped transcripts must initially be dormant. In other words, transcripts only become active if they arrive at a marked terminal. Si thought that one way cells transform dormant mRNA transcripts into active proteins at marked synapses would be to recruit a regulator of mRNA translation.

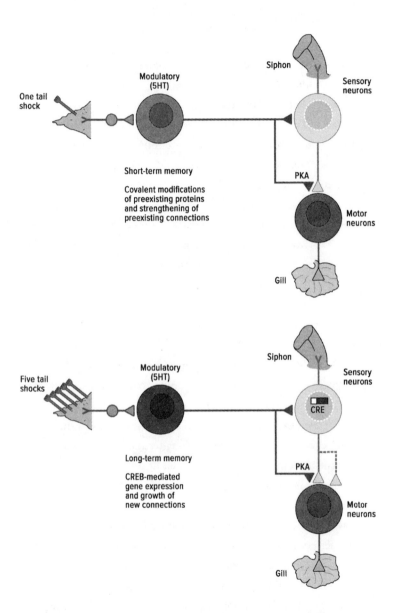

FIGURE 2.1 Molecular mechanisms of memory. Short-term memory involves strengthening existing connections between neurons (above); long-term memory requires the growth of new connections (below).
Artist: Sarah Mack

The oocytes of *Xenopus*, an aquatic frog, provide an example of such regulation. In this frog, the messenger RNA is silent until activated by a regulator of protein synthesis known as CPEB, the cytoplasmic polyadenylation element binding protein.[2] Si searched for a homolog of CPEB in *Aplysia* and found that in addition to a form of the protein that is important for development, there was a new form of the protein that had novel properties.[3] Blocking this new form of CPEB at a marked (active) synapse *prevented* the maintenance of long-term memory. This raised the question: How does this new form of CPEB contribute to the maintenance of long-term memory?

A remarkable feature of the *Aplysia* CPEB is that one region (its N-terminus) resembles the domain that is characteristic of prions. Prions are misfolded proteins that perpetuate their misfolded state by causing other, normal proteins to misfold; these prions then form clumps inside the neuron, disrupting its function and eventually killing the cell. In this way, prions give rise to neurological diseases in humans and in experimental animals. The *Aplysia* CPEB is endowed with similar self-sustaining properties, but unlike other prions, it appears to be beneficial: that is, the active, self-perpetuating form of the protein does not kill cells, but rather controls protein synthesis at marked synapses. Notably, the persistence of long-term memory in *Aplysia*, as well as in the fruit fly *Drosophila* and in mice, depends on a functional CPEB prion element.[4]

Based on these findings, Si and I proposed a model of memory storage in *Aplysia* that relies on the prionlike properties of CPEB.[5] We based our model on two characteristics of CPEB: 1) it activates the translation of dormant mRNA transcripts, and 2) it exists in two states—one that is active and one that is inactive and actually represses mRNA translation. In an unmarked synapse, the level of

CPEB is low and the protein is inactive. According to our model, serotonin—a neurotransmitter that is released during learning tasks in *Aplysia*—triggers an increase in the synthesis of CPEB. If a given threshold of CPEB production is reached, CPEB is converted to the prionlike state, which is more active and unable to inhibit mRNA translation (fig. 2.2).

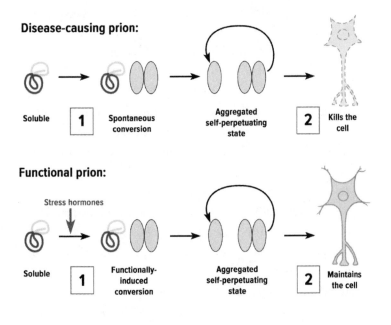

FIGURE 2.2 Comparison of disease-causing and functional prion states. Disease-causing prions undergo a spontaneous conversion to the prion state (above). Stress hormones released in response to an electric stimulus in learning tasks convert CPEB to the prion-like state (below).
Artist: Sarah Mack

Once the prionlike state is established at a marked synapse, dormant mRNA transcripts, made in the cell body and distributed throughout the cell, are translated into protein—but only at that marked synapse. The prionlike state helps to maintain the long-term, synapse-specific changes required for memory storage in two ways: 1) it stabilizes the new synaptic growth induced by learning, and 2) it makes possible long-term memory storage during periods of normal growth, when very low levels of protein synthesis are required.

UNCOVERING THE BIOLOGICAL BASIS OF THE GATEWAY EFFECT IN DRUG ADDICTION

In a paper published in *Science* in 1975, my wife, Denise, described her pioneering work on the epidemiology of the *gateway effect*—the finding that young people become involved in drug use in a well-defined sequence of stages.[6] She had observed that in human populations, drug usage often starts with the consumption of a legal drug and only then proceeds to illegal drugs: that is, the use of cigarettes and alcohol generally precedes the use of marijuana, which in turn precedes the use of cocaine or other illicit drugs. Since 1975, additional epidemiological studies have found that in human populations nicotine use is a gateway to cocaine use, thus increasing the risk of addiction to cocaine among people who smoke.

What was not clear from these epidemiological studies is *how* nicotine exerts its gateway effect in the body. In 2003 Denise wrote an editorial in the *Journal of the American Medical Association* calling attention to the need for collaboration between epidemiologists and scientists who study the neurobiological mechanisms underlying behavior in animal models.[7] She was also looking for collaborators who would be able to design and implement such interdisciplinary experiments.

By this time, scientists had learned that some of the same cell-biological processes involved in memory are also involved in addiction, and that the stepwise process of the gateway effect has interesting similarities to the stages of addiction. I thought this knowledge put us in a position to look seriously at the molecular and cellular mechanisms that might underlie the gateway effect. As it happened, I had been invited to give a basic science lecture at the international meeting of the Society for Research on Nicotine and Tobacco, a meeting that Denise was attending. After the lecture, Denise and I talked and realized we could move forward together in planning experiments that would test the gateway effect. So in 2010 Amir Levine and I collaborated with Denise to bring the techniques of molecular biology to bear on this epidemiological question.

We began by developing a mouse model of drug use and then used it to study the action of nicotine in the brain and to examine the effects of nicotine on the animal's subsequent responses to cocaine. We found that giving mice nicotine increased their response to cocaine, as assessed by an increase in addiction-related behaviors and by reduced long-term potentiation in the nucleus accumbens, a part of the striatum. Long-term potentiation strengthens the connections between synapses, and this strengthening underlies a number of forms of learning and memory. The striatum is part of the brain's reward system, which regulates the production of the neurotransmitter dopamine in response to pleasurable stimuli and the anticipation of reward. Our brain's normal attraction to pleasure and the anticipation of reward—the memory of pleasure—can be subverted by drugs. As a result, the connections between synapses become weaker, causing the brain to produce an excess of dopamine, leading eventually to addiction.

Moreover, we found that the increased response to cocaine occurred only when nicotine was administered *before* the concurrent administration of nicotine and cocaine. Reversing the order of administration revealed that cocaine has no effect on nicotine-induced behaviors or on long-term potentiation in the striatum.

We inferred from this finding that nicotine primes the mouse's response to cocaine by enhancing cocaine's ability to induce expression of a gene called FosB. Previous studies had found that increasing expression of the FosB gene in the striatum of mice results in depressed long-term potentiation and excessive dopamine production in that region of the brain—conditions that pave the way to addiction. Our inference was borne out in further tests with a compound that simulates the action of nicotine; this compound also primes the mouse's response to cocaine and enhances expression of the FosB gene.

The epidemiological data indicate that most cocaine users started using cocaine after they had begun to smoke and while they were still smoking, and that initiating cocaine use after they had started smoking increased the risk of becoming addicted to cocaine. These data in people are consistent with our molecular findings in mice. If the findings in mice apply to people, a decrease in smoking rates among young people can be expected to lead to a decrease in cocaine addiction. Moreover, we also explored the effect of prior alcohol use on the future use of cocaine and, remarkably, observed a similar gateway effect. That is, similar to nicotine, alcohol potentiates the effects of cocaine.[8]

These findings about nicotine, initially published in *Science Translational Medicine* in 2011,[9] were novel and therefore very encouraging. In addition, they showed clearly that molecular neuroscience can provide insight into a public health issue. Our successful collaboration was also personally gratifying, as it helped allay Denise's fears that a post-Nobel intellectual decline is a certainty. Based on this collaboration, Denise and I were invited to give the Shattuck Lecture at the annual meeting of the Massachusetts Medical Society in 2014.[10]

By the time of the lecture, another question regarding addiction had arisen: the effect of using electronic cigarettes (e-cigarettes, or vaping). Many people thought e-cigarettes were an important public health boon because delivering nicotine via a vapor rather than smoke does not cause lung cancer. But, as Denise pointed out, the effects of nicotine on the brain are every bit as strong; in fact, e-cigarettes are a perfect delivery system for nicotine to the brain. Moreover, the flavorings added to the vapor might themselves prove harmful.

These concerns indicated that for older adults who have been smoking a long time, or who want to give up smoking, e-cigarettes can be beneficial in reducing the risk of cancer. However, the situation is different for young people: e-cigarettes are unlikely to reduce the role of nicotine as a gateway drug to addiction to other substances. In addition, while cigarette smoking is at a record low among teenagers, vaping has increased sharply in recent years. Because of their serious health risks, the U.S. Food and Drug Administration has banned the sale of vaping devices, liquid nicotine, and all flavored e-cigarette products to anyone under age 21.

IDENTIFYING DOPAMINE'S ROLE IN THE
COGNITIVE SYMPTOMS OF SCHIZOPHRENIA

The presence of excessive dopamine in some regions of the brain contributes to several of the symptoms of schizophrenia, a disorder that affects about three million people in the United States. Schizophrenia produces wide-ranging effects on cognitive function and behavior, making the biology of the disorder particularly difficult to sort out. However, recent advances in genetics and in brain imaging have given us new insights into the biological basis of schizophrenia.

As a first step, Eleanor Simpson, Christoph Kellendonk, and I developed a mouse model of the cognitive and negative symptoms of schizophrenia.[11] In people, the cognitive symptoms include problems with volition, executive function, and working memory (a form of short-term memory), as well as other cognitive features reminiscent of dementia. Negative symptoms include social withdrawal and listlessness. Positive symptoms of schizophrenia, in contrast, reflect disordered thinking, which detaches a person from reality, leading to psychotic behavior such as hallucinations and delusions. Each category of symptoms results from disturbances in a different part of the brain.

Simpson, Kellendonk, and I knew that antipsychotic drugs, which are very effective for treating the positive symptoms of hallucinations and delusions, produce their effects by blocking dopamine receptors in a particular neural pathway in the brain, thus attenuating the action of dopamine there. In addition, brain-imaging studies had revealed that people with schizophrenia have more dopamine and, importantly, more of a particular class of dopamine receptor—the D2 receptor—in the striatum. The D2 receptor was a particularly interesting target because in some people the increased number of receptors may be determined genetically.

In light of these findings, we set out to determine whether an excessive number of D2 receptors in the striatum causes the cognitive symptoms of schizophrenia. To do so, we created a mouse model containing a human gene that overexpresses D2 receptors in the striatum; we could reverse the overexpression of D2 receptors by switching the gene on and off.

We found that mice with this translocated human gene, or transgene, exhibit cognitive impairment in tasks requiring behavioral flexibility and in tasks requiring working memory, but they do not exhibit more general cognitive deficits. In addition, the mice lack motivation, a characteristic of the negative symptoms of schizophrenia. The most interesting finding was that when we switched the transgene off, the motivational deficits disappeared but the cognitive deficits did not—they persisted long afterward. We therefore concluded that the cognitive deficits we observed result not from continued overexpression of D2 receptors, but from excess expression during prenatal development.

To determine what may mediate these observed cognitive deficits, we analyzed activity in the prefrontal cortex, the brain structure mainly associated with working memory. We found that overexpression of D2 receptors in the striatum affects three physiological measures in the prefrontal cortex: the concentrations of dopamine, the rates of dopamine turnover, and the activation of D1 receptors. All three of these measures are critical for working memory.

AMELIORATING AGE-RELATED MEMORY LOSS

Our memory generally begins to weaken around age 40. Whether that age-related memory loss is benign or signals the onset of Alzheimer's disease is a matter of great concern—and not just to individuals. It has enormous financial and emotional consequences for our society, especially given our aging population.

To address the question of whether age-related memory loss is distinct from Alzheimer's disease, Scott Small, Elias Papadopoulos, and I set out to distinguish the molecular underpinnings of these memory disorders. We knew that Alzheimer's disease begins in the entorhinal cortex, a region that interfaces with the hippocampus and is relatively unaffected by aging. Earlier, Small and his colleagues had found that age-related memory loss involves the dentate gyrus, a neighboring region of the hippocampus that is activated by the entorhinal cortex.[12]

We first carried out a gene expression study in postmortem tissue obtained from people between the ages of 40 and 90 who did not have Alzheimer's disease (fig. 2.3). We identified 19 genes that exhibit reliable age-related changes in expression in the dentate gyrus. The most striking change we observed as individuals grew older was a decline in the expression of the RbAp48 gene, which codes for a protein that modifies the acetylation of histones and thereby regulates the expression of many genes. We observed no decline in expression of the RbAp48 gene in the entorhinal cortex.

To test whether the decline in RbAp48 protein that we had observed in human tissue could be responsible for age-related memory loss, we turned to genetic studies in mice. Those studies revealed that, consistent with humans, the RbAp48 protein is less

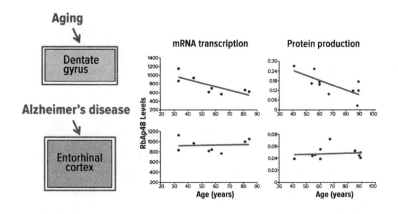

Elias Pavlopoulos et al. 2013

FIGURE 2.3 Gene expression and memory loss. Expression of the RbAp48 gene as measured by mRNA transcription (center) and protein production (right) decreases linearly in the aging human dentate gyrus (green) but not in the entorhinal cortex (red) or other regions of the hippocampus involved in Alzheimer's disease.
Artist: Sarah Mack

abundant in the dentate gyrus of old mice than in young mice (fig. 2.4, left).

We next developed a genetically modified mouse that expressed an inhibitor of the RbAp48 protein in its forebrain when the animal reached adulthood. When we inhibited that protein in young mice, it caused memory deficits similar to those associated with aging (fig. 2.4, center). Functional magnetic resonance imaging (fMRI) of the hippocampus showed that the dysfunction occurred only in the dentate gyrus of the mice and that it corresponded to

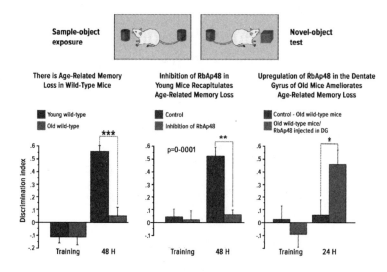

FIGURE 2.4 RbAp48 protein and memory loss. Inhibiting expression of the RbAp48 gene, and thus the amount of RbAp48 protein, in the dentate gyrus of young mice causes memory deficits characteristic of old mice, as indicated in spatial object recognition tasks. Increasing the amount of protein in the dentate gyrus of old wild-type (non-genetically modified) mice ameliorates age-related memory loss.
Artist: Sarah Mack

a decrease in histone acetylation in this region of the brain. Increasing the amount of RbAp48 protein in the dentate gyrus of old wild-type (non-genetically modified) mice ameliorated the abnormalities in histone acetylation and therefore the age-related memory loss (fig. 2.4, right).

Together, these findings showed that normal age-related memory loss takes place in the dentate gyrus and that a specific molecule associated with cognitive aging, the RbAp48 protein, could be modified to reduce memory loss. Figure 2.4 summarizes the clearest evidence to date that age-related memory loss and Alzheimer's disease are distinct disorders.

CHAPTER 3

ADVENTURES IN THE PUBLIC UNDERSTANDING OF SCIENCE

B eginning in high school, I have always taken a great deal of pleasure from writing. My interest in writing about science began later, in 1968, when Alden Spencer and I were invited to write a review, "Cellular Neurophysiological Approaches in the Study of Learning," for *Physiological Reviews*.[1] Several years later, in 1976, after becoming fascinated with the function of the hippocampus, a structure that exists only in the brain of vertebrate animals, I wrote my first book, *Cellular Basis of Behavior: An Introduction to Behavioral Neurobiology*.

When I came to Columbia University in 1974 to develop the neuroscience curriculum, I was struck by how much energy students were devoting to writing out the lectures that my colleagues and I delivered. I wanted to help them get over that, so I encouraged the faculty to provide a syllabus for each lecture. In time, I edited them into one document, added figures to it, and generally improved it. In 1981 my colleague James H. Schwartz and I decided that the syllabus was becoming sufficiently useful that we might make a textbook out of it. That textbook, *Principles of Neural*

Science, was the first attempt to link cell and molecular biology to neural science and to link neural science to behavior and to clinical neurology and psychiatry.

The response to the first edition was so gratifying that we worked on a second edition, attempting to make the book better and more complete. With the second edition, not only students but also scientists began to regard our textbook as useful. With the help of our colleague Thomas Jessell, we further improved the third and fourth editions. Now entering its sixth edition, *Principles of Neural Science* is the most widely used neuroscience text in the world. It has been a source of deep satisfaction to me and to the many other Columbia faculty who have contributed to it.

Writing this textbook gave me an opportunity to reach out to a less specialized audience than the scientists in my field—namely, medical students and graduate students. It also led to my giving a number of invited lectures for the general public. My interactions with those audiences and their positive response to the lectures convinced me that I could explain the key issues of brain science to nonscientists.

The Nobel Prize gave me an opportunity to do this and inspired me to write a book focusing on my own research, my first book for a general audience. Since then, I have greatly enjoyed contributing to the public understanding of science through other books and essays, talks, and television programs, explaining the new findings in neuroscience, the thinking and research studies that led to them, and what those findings tell us about ourselves.

In the process, I have learned that the general public is eager to understand how the physical workings of the brain give rise to our mind, our emotions, and our behavior. People want to know how the brain may become disordered in disease and how those disorders may be ameliorated. Finally, they want to know how our knowledge about the brain can be used to improve both individual lives and public policy.

IN SEARCH OF MEMORY: THE EMERGENCE OF A NEW SCIENCE OF MIND

A person who is awarded the Nobel Prize is asked to do two things in preparation for the visit to Stockholm: prepare an autobiographical sketch and prepare a lecture. So one or two days after I learned that I had won the prize, Harold Varmus, who had won a Nobel Prize several years earlier for his work on oncogenes and who had moved to New York to serve as head of the Memorial Sloan Kettering Cancer Center, called me up and said, "I've got to speak to you."

I literally ran over to see him. "Look," he said. "This autobiographical sketch that you are asked to write is important. Take it seriously. A lot of people just hand in a CV. You know how to do this; write about your life. Start now. Take the time to do it right."

I knew I had to prepare a lecture and that it would take a great deal of work, so I was obviously prepared to put time into it. But I had not given much thought to the autobiographical sketch. Since Harold thought it would be very worthwhile for me to actually write a full-length essay, I became determined to do just that.

So I wrote an essay about my life as a Jewish child in Vienna, my experience when Hitler marched into Vienna and was received with enormous enthusiasm by the Viennese, and how my former classmates, almost all of whom were non-Jewish, turned on me. I described how my family was expelled from our apartment during

Kristallnacht and how we ultimately fled to the United States, where we have since had a marvelous life. I had not told most of my friends any of these things about myself, so when they read the essay, they were both astonished by the facts and enthusiastic about me writing about my life story.

The Nobel Foundation publishes a volume, *Les Prix Nobel,* which contains the autobiographical sketch and lecture of each winner of that year's prize. My essay was called "The Molecular Biology of Memory Storage: A Dialog Between Genes and Synapses." I saw from the response of my friends that they really enjoyed reading it, and that gave me the idea of putting the lecture and the autobiographical essay together and making a book out of it, a scientific autobiography. I began writing it in 2004.

The result was *In Search of Memory: The Emergence of a New Science of Mind.* In it I describe my childhood in Nazi-occupied Vienna; my family's flight to America; and my education, first as a historian seeking to find some rationale for why the Viennese turned so suddenly and viciously against the Jews, then as a physician and psychiatrist, and ultimately as a brain scientist. I then chronicle my search for the biological basis of learning and of how memory is formed and stored in the brain. Along the way I wrote about my family, including Denise's experiences as a child in occupied France; my friends and colleagues; and finally about my tentative reconciliation with Austria.

In the process of describing my work on memory, the book reveals one of the key strategies scientists use in their research: seeking to understand complex processes by reducing them to their essential actions and studying the interplay of those actions. In my case, I began with an important question that fascinated me: How is memory stored in the brain? I studied it first in a simple system (the marine snail *Aplysia*) and then in a more complex one (the hippocampus of mice).

Learning and memory are two of the most remarkable capabilities of our mind. Learning is the biological process of acquiring new knowledge about the world, and memory is the process of retaining and recalling that knowledge over time. Most of our knowledge of the world and most of our skills are not innate; they are learned. Thus we are who we are because of what we have learned and what we remember—and also because of what we forget. Without the unifying power of memory, our life would be fragmented into a meaningless series of events. Likewise, if we remembered everything that happened to us, our life would be burdened by irrelevant details. The selective process of memory holds our mental life together and provides it with structure.

During the last decades of the twentieth century, the most valuable insights into the human mind came from the merger of a philosophy of mind with cognitive psychology, the science of mind, and the subsequent merger of those fields with neuroscience, the science of the brain. This synthesis was energized by dramatic achievements in molecular biology, brain imaging, and theoretical neuroscience. The result was a new science of mind that extends from the molecular biology of the brain to the systems biology of the brain and then to behavioral psychology. Within this new, interdisciplinary structure, the scope of memory research ranges now from genes to cognition, from molecules to mind.

The new science of mind is based on four principles. First, our mind and our brain are one. The brain constructs our mind. It constructs our perceptions of the world, regulates our thoughts and emotions, and controls our actions. The brain is responsible not only for relatively simple motor behaviors, such as running and eating, but also for complex acts such as thinking, speaking, and creating works of art. Looked at from this perspective, our mind is a set of operations carried out by the brain.

Second, all functions of the brain, from the simplest reflex to the most complex, creative endeavors, result from the activity of specialized neural circuits, and all of those circuits are made up of the same elementary signaling units: the neurons, or nerve cells.

Third, neural circuits use specific molecules to generate signals within and between nerve cells.

Fourth, those specific signaling molecules have been conserved through millions of years of evolution. In other words, we use some of the same molecules to navigate through our environment as our most ancient and distant ancestors did—organisms such as single-celled bacteria and yeast, or simple multicellular worms, flies, and snails.

Thus, the new science of mind gives us not only insights into ourselves—how we perceive, learn, remember, feel, and act—but also a new perspective on ourselves in the context of biological evolution. It makes us appreciate that the human mind evolved from molecules used by our simpler ancestors and that the extraordinary conservation of the molecular mechanisms that regulate life's various processes also extends to the life of the mind.

In Search of Memory was published in 2006 and went on to win major awards, including the 2006 *Los Angeles Times* Book Prize for Science and Technology and the 2007 National Academies Best Book Award for Excellence in Communicating Science to the General Public.

THE BRAIN SERIES

Shortly after winning the Nobel Prize I was invited by the television commentator Charlie Rose to appear on his PBS program, which I did in August 2001. We had an extremely good discussion, and later, when *In Search of Memory* came out, he invited me to join him on his program again, to discuss the book.

When the date for that discussion arrived, Charlie, who was in Europe at the time, called his office to say that he had to undergo an emergency medical procedure but that he would like the program to continue with Harold Varmus interviewing me, if that were fine with me. Of course, it was perfectly fine with me. So I had a wonderful session with Harold, discussing my book.

Some weeks later, when Charlie had returned, he invited me to an annual conference in Aspen, Colorado, where he chaired a series of roundtable discussions. He asked me to organize a discussion on aging and the brain, focusing on whatever topic I wanted. I chose age-related memory loss. The other roundtables at the conference were on Iran, on Afghanistan, on Israeli-Palestinian interactions— one depressing world crisis after another. So as the audience moved from one roundtable to the next, they became progressively more despondent. When the audience came to our roundtable, they found it was filled with biologists who were optimistic about the progress being made in our understanding of age-related memory loss and how it might be ameliorated. The discussion was excellent, and the audience loved it.

Following the success of the roundtable at Aspen, Charlie asked me in 2008 to help him start a television series about brain science. I was to be the co-host and to help organize the topics, as well as invite the guest panelists. The result was the Charlie Rose *Brain*

Series, which began running on PBS in 2009. In time, there were three *Brain Series,* which ran from 2009 to 2017.[2]

Charlie introduced each program in general terms, and I followed with a slightly more technical introduction to the topic at hand. The panelists I selected for these programs were all accomplished, articulate people whose work and experiences with various aspects of brain science and brain disorders contribute greatly to our understanding of the human mind, brain, and behavior. These television series provided me with an opportunity that I had not anticipated—the privilege of conveying to a wide, general audience the astonishing progress that has been made in brain science and its effects on society.

1. An Overview of the Brain

The first *Brain Series* ran for a year, with one program a month, and provided a general overview of the brain. Each program featured four outstanding panelists in a roundtable format. I saw the series as a wonderful opportunity to move education about the biology of the brain in a new direction, using the excitement and spontaneity of roundtable discussions to communicate complex topics in straightforward, simple terms that everyone could understand.

In the first ten programs, we discussed how various human mental processes arise from the brain. We explored the mysteries of the brain—consciousness, free will, perception, cognition, and memory—in the first program. We then moved on to discussions of the visual brain, the acting brain, the social brain, the developing brain, the aging brain, the emotional brain, the fearful and anxious brain, the mentally ill brain, and the neurologically disordered brain. Each of the programs began with an explanation of the basic biology of a particular mental function and then focused on one or more diseases associated with that mental function.

In the last two programs, on the deciding brain and the creative brain, we began to look at the extent to which neuroscience can inform other bodies of knowledge. We saw how insights into the biology of decision making can increase our understanding of how we make economic decisions, and even moral decisions. The final program explored the most complex human activity, creativity. Our brain has enormous creative capabilities, and although we see creativity in exquisitely refined form in works of art, it is evident in every aspect of life.

As it turned out, the roundtable discussion format enabled us to provide a systematic view of how the brain gives rise to mind, of how the various circuits of the brain account for our ability to perceive, to feel, to think, and to act. We explored how these neural circuits underlie all aspects of human cognitive functioning and social interaction, as well as how our knowledge about those circuits gives us insights into the diseases that haunt humankind, such as schizophrenia, depression, anxiety states, and post-traumatic stress disorders.

Consciousness

A wonderful aspect of neuroscience is that it is a natural bridge between the humanities and the sciences. One example of this—and a subject of our first program as well as a later one—is consciousness. Consciousness has three remarkable features. The first is qualitative feeling: listening to music is different from smelling a lemon. The second is subjectivity: I know you are feeling pain when you burn your hand, but not because I am actually feeling your pain—only you can feel that. I feel pain only when I burn myself. The third feature is unity of experience: I experience the feeling of the breeze on my skin and the sound of my voice and the sight of the other people sitting around the dining table as part of a single, unified consciousness—as my experience.

How does the brain produce consciousness? This question has intrigued philosophers and brain scientists alike. Brain imaging can reveal correlations between the experiences that people have and the areas of the brain that are active when they have those experiences. But we now need to explore how the brain goes from neurons firing to generating feelings that allow us to experience the world as a unified whole rather than simply a collection of perceptions, to our sense of self. The question of how the brain produces consciousness is arguably the most difficult problem in all of science.

Decision Making

Decision making is another aspect of our lives in which neuroscience has acted as a bridge between the sciences and the humanities. In fact, decision making in economics and in moral philosophy are perhaps the areas in which neuroscience has had the most substantial influence. Early biological studies showed that even though decision making has a variety of facets—perceptual, personal, social, economic, moral—all decisions, whether simple or complex, involve the same key areas of the brain. Moreover, while we generally prize our ability to make rational decisions, no decision is purely rational. If we can understand the principles underlying decision making, we can predict behavior more accurately.

Unconscious emotion is a factor in every decision we make. Moreover, that emotional response is largely determined by the way the decision is framed. This is true in both economic and moral decision making. A famous gambling experiment devised by Daniel Kahneman and Amos Tversky illustrates the importance of how a decision is framed. If I give you $50 and tell you to choose between making a bet in which you keep $20 and making a bet in which you lose $30, it doesn't matter which bet you choose: the outcome will be the same. But people choose one bet or the other depending on whether the choice is framed positively (keeping $20) or negatively (losing $30).

This example of framing exposes the essential flaw in the economic model commonly used to explain choice: namely, economic rationality. That model assumes people will make the choice that results in the best long-term economic outcome; however, as we saw in the gambling experiment, people do not choose on that basis. Many other experiments have confirmed that the way we convey information, how we frame choices, is critically important, both in clinical settings and in public policy.

Neuroscience has also been applied to moral decision making—and it shows that these decisions, too, involve emotion as well as rationality. We can categorize moral decisions as impersonal or personal on the basis of which areas of the brain they activate. Impersonal moral decisions are shaped by conscious decision-making processes, in which a network of brain regions assesses the various alternatives, then sends the verdict onward to the prefrontal cortex, where a person chooses the superior option. Personal moral decisions are shaped by our unconscious emotions. In these decisions, a network of brain regions associated with the processing of emotions is activated. Interestingly, the difference between personal and impersonal moral decisions is built into our brain, regardless of what culture we live in or what religion we subscribe to. The two types of moral decisions trigger different patterns of activity in the brain.

Criminal courts have an acute need to understand the value and limitations of neuroscientific evidence when making decisions. In 2012 the U.S. Supreme Court ruled that a sentence of life in prison without parole for juvenile criminals is unconstitutional. Neuroscience was central to that decision: the justices pointed to scientific findings that adults and adolescents differ in "parts of the brain involved in behavior control." While that decision relied on the appropriate use of neuroscientific evidence, other cases have raised more tenuous uses. In a recent murder trial in Maryland, the judge ruled that functional MRI scans are not—at least not yet—reliable lie detectors. Expert witnesses at that trial offered differing views of the scans' usefulness in a real-world context.

Judges are looking to psychologists and neuroscientists for help in interpreting neurological findings. They want to know if the findings are reliable, what they mean in terms of behavior, and how they should be used in the law to improve the fairness of the judicial system.

2. The Patient Speaks

The first *Brain Series* was extremely popular and led Charlie and me to undertake a second year-long series. This second series, called

"The Patient Speaks," focused on disorders of the brain and included people with specific disorders of cognition, movement, mood, consciousness, and social interaction. I particularly wanted this series to focus on patients because I thought that the experiences and insights of people living with a disorder would help the audience understand the difficulties that they and their families face and the strength and courage needed to cope with the disorders.

The second series includes the most common brain disorders: schizophrenia, addiction, agnosias, neurodegenerative diseases such as Alzheimer's, autism, depression, motor disorders, amyotrophic lateral sclerosis (ALS), multiple sclerosis, pain, post-traumatic stress disorder, hearing loss, and blindness. In the past, these brain disorders would have been separated into two categories: psychiatric disorders and neurological disorders. The categories were considered to be fundamentally different: neurological disorders were viewed as affecting the brain, whereas psychiatric disorders were viewed as affecting the mind. Moreover, in the days before sophisticated brain-imaging technology, psychiatric disorders could not be traced to any physical damage in the brain. This led to the stigmatizing of mental patients, who were viewed as essentially weak-willed or defective.

Today, scientists realize that the brain and mind are inseparable. As a result, our understanding of brain disorders is more nuanced. Both neurological and psychiatric disorders affect aspects of brain function: perception, action, memory, volition, motivation, emotion, empathy, social interaction, thought, attention, creativity, sexuality, and consciousness. In fact, the two classes of disorders overlap extensively at the margins. For example, Alzheimer's disease, which primarily affects memory, and Parkinson's and Huntington's disease, which primarily affect movement, are all thought to involve a particular mechanism in the brain known as protein misfolding. The three disorders produce strikingly different symptoms because the abnormal folding affects different proteins and different regions of the brain. We are beginning to discover common mechanisms in other diseases as well.

As I expected, the patients in this second series provided invaluable insights into their disorders. For example, a noted artist with prosopagnosia, the inability to recognize faces, explained that he paints portraits in order to understand the faces of the people he knows and loves, to fix them in his memory. A young woman with autism pointed out that her preoccupation with details and patterns is actually a strength in her job—a job that she does well. A man with schizophrenia spoke to the stigma still surrounding this disorder. He wanted to be able to hold a conversation with somebody who would respond to him as a person with a neurological disorder, not a crazy person. He added that he would like to see a national discourse about schizophrenia similar to those surrounding autism and breast cancer.

According to PBS, more than 260,000 households, on average, watched the first and second *Brain Series* each week. This overwhelmingly positive response showed us that many people with no background in science were eager to learn about the new science of mind—and it led to a third series.

3. Brain Science and Society

The third series, which was not finished, began to explore some of the critical social issues that are directly influenced by brain science: aggression and violence, parenting, gender identity, sports-induced concussions, and childhood trauma. Each program focused on one issue and provided a biological basis for public policy decisions regarding that issue. All of the issues we covered in the third series have important, widespread public policy implications for people of all ages, but some of them, including brain injuries and gender identity, are particularly salient for children and adolescents.

Physical Injuries to the Brain

Brain injuries caused by sports are among the leading causes of death and disability in young people in the United States. Moreover, our increasing understanding of the neurological effects of traumatic brain injuries has revealed that young people are particularly susceptible to them. These facts have ignited a controversy about the risk of contact sports, which has emerged as one of the major health issues of our time.

As a society, we generally believe that team sports are advantageous for young people, both physically and socially. In the United States, 50 million young people play team sports. They learn early on that frequent, regular exercise is healthful, and many of them are likely to continue to do exercises of one sort or another for the rest of their lives, ensuring them a healthier life span than they would have enjoyed otherwise. Team sports also instill important social values: a sense of honesty, fair play, and teamwork. These very beneficial values also help carry young people through a lifetime. For some young athletes, team sports lead to college scholarships and perhaps even to professional careers. But we have become aware that at the very highest level—professional football, soccer, and hockey, for example—team sports are associated with severe, chronic brain damage, and that has made us worry whether we are protecting our young people sufficiently.

The peak ages of injury, for both boys and girls, are about 12 to 19. Teenagers are more susceptible to brain injury than adults because of the important developmental changes in the brain that take place during adolescence. We do not yet have any drugs or surgical interventions to treat traumatic brain injury, so it is incumbent upon everyone involved in team sports—coaches, trainers, teachers, parents, and the players themselves—to recognize the signs of a concussion and to realize that if a concussion may have occurred, the athlete must immediately be pulled from the game and not allowed to play again until cleared by a health professional. This is extremely important, both because an adolescent who has had one concussion is much more likely to have a second concussion and because it will take longer to recover after a second concussion. In addition, someone who has had two concussions is much more likely to have a third or fourth one.

Although we have some understanding of how concussion affects the brain, we need additional neuroscience research, additional data, better safety equipment, and better sports protocols. It is critical that more young athletes understand how important it is to prevent injuries and to admit it when they do have one. This is an enormously important change in sports culture, which has, in the past, emphasized getting back on the field for the team and "toughing it out" after being injured.

Psychological and Social Injuries to the Brain

Physical injuries, such as concussions, are not the only source of trauma to the brain. Psychological trauma can also cause permanent brain damage. This is true for people of any age, as post-traumatic stress shows us, but trauma that occurs early in life, while the brain is developing, can be particularly damaging. In some cases, early social and psychological adversities such as loss of a parent, parental abuse, parental neglect, poverty, or bullying can cause more severe damage to the brain than a physical injury. While there are many examples of psychological and social traumas in children, I focus here on two of them: institutionalization and family poverty.

Institutionalization of Children. In the early 1940s, a psychoanalyst named Rene Spitz carried out a remarkable study of children isolated from their mothers at birth. The children lived in two very different environments. One environment was a nursery connected with a prison, for children whose mothers had given birth to them while incarcerated. The other was a foundling home, for children who had been abandoned by their mothers. In the prison nursery, the mothers were allowed to interact with their infants at certain times during the day. As a result, these mothers were able to bestow a lot of affection on their children. In the foundling home, the infants were assigned to a nurse. Each nurse took care of seven infants. As a result, the infants received a limited amount of attention and lived in a situation of relative sensory and social deprivation.

When children from both environments were examined a year later, the differences between them were readily apparent. The children in the nursery, who had had contacts with their mothers, were happy and interacted well with the people around them. The children in the foundling home were anxious and not very curious about their surroundings. At ages two and three the differences were even more dramatic. The children who were in the prison nursery walked and talked and were very gregarious in interacting with one another. In contrast, most of the children in the foundling home could not walk, most of them could not talk, and those who did talk could only express themselves with a few words.

Subsequent studies, including recent brain-imaging studies of children in orphanages in Romania by Charles Nelson, a developmental cognitive neuroscientist, have confirmed these findings. They show that the first two years of life are a critical period of development. Children deprived of loving social interaction with a parent or other caregiver during their first two years of life have less electrical activity in the brain. The result is alterations in brain function and structure that affect both cognition and communication between different areas of the brain. These alterations can result in lower-than-normal IQ, dramatic increases in mental health problems, and changes in memory. A comparable period of isolation later in life has very little effect.

Can anything be done to avoid these negative effects on the brain of a child? Nelson's study of orphaned children in Romania indicates that children removed from an institution and placed with a good foster family before the age of two attain the same level of brain function by about age four as children who were never institutionalized. The outcomes for children placed in foster homes after the age of two were less reassuring; in fact, brain activity in children placed after age two was the same as that of children who had remained in the orphanage. However, other studies of older children placed in a good home have shown that these children, too, can benefit from the change. In any case, the sooner a child can be moved out of a deprived environment and into a good environment, the easier it is for them to attain healthy brain function.

Family Poverty. Living in poverty is a source of psychological and social trauma for people at any age, but particularly for children and adolescents, whose brains are still developing. Studies have found that school-age children from socioeconomically advantaged families, in general, outperform children from disadvantaged homes on memory tasks. Does that disparity in performance stem from a neurological difference or social differences?

Scientists have known since the 1950s that the hippocampus, located deep within the brain, is critical for memory storage. They have also known for some time that in many cases, a larger hippocampus is associated with greater memory skills. So they used brain imaging to compare children from homes with greater family incomes with children from economically disadvantaged homes. They found that generally, children from more advantaged environments had a larger hippocampus.

This is an absolutely amazing result—to find that poverty affects the size of the hippocampus and therefore a child's cognitive and memory capabilities. But this finding raises the question of *why* children from disadvantaged families generally have a smaller hippocampus.

One likely reason is stress. We know that in families facing economic hardship, the children as well as the adults must deal with a host of stressors every day. We have also known for quite some time that stress has profound and cascading effects on the development of the hippocampus.

Often, the stress that living in poverty exerts on parents—worry about providing food, clothing, and shelter; keeping their children safe; spending long hours away from home working two or more jobs; uncertain job security—results in parents having little free time to interact with their children. A study of children brought up in an environment lacking appropriate social bonding with parents and appropriate intellectual stimulation has shown that certain pathways in the brains of these children fail to develop satisfactorily. As a result, the hippocampus may not reach its normal size.

Adults who experience such deprivation early in life are also more susceptible to depression and to suicidal thoughts. They are more likely to use drugs and more likely to develop metabolic and cardiovascular diseases. Even moderate impairment in the bonding between parents and children can have a negative effect.

How does this physical effect on the developing hippocampus come about? It does so through alteration in gene expression.

Normally, an infant starts with a blank slate, genetically speaking. Over its lifetime, the infant's experiences add and subtract transcription factors that activate genes. But as we have learned from studies of mice, depriving an infant mouse of access to its mother will cause alterations in gene expression—alterations that can be maintained not just for the life of that infant mouse but for several generations thereafter. So in addition to its inherited genes, the mouse has epigenetic traits, or changes in gene expression, that are then passed on from one generation to another. These findings may be applicable to humans, but the evidence thus far is not conclusive.

What is certain is that lack of strong parental bonding and the social and emotional benefits that flow from that bonding can have numerous serious effects on children. It can lead to changes in DNA that increase children's vulnerability to stress, alter the development of the brain, produce unhealthy responses to stress, and give rise to cognitive health consequences as well as social consequences. Research points to a link between these adverse consequences and the stress associated with poverty. Perhaps if we can develop social programs that reduce family poverty—and thus alleviate the stress it causes—we will be able to improve these children's cognition and memory.

Gender Identity

Gender identity is another area in which neuroscience and public policy intersect in a way that can profoundly affect the lives of children and adolescents. Gender identity is our sense of where we belong on the continuum of sexuality, of being male, or female, or neither, or both. It begins early in childhood and is not based simply on anatomical sex. This is why a child can feel trapped in the wrong body, being expected to behave in certain ways but feeling and wanting to behave differently.

For children whose gender identity is different from their anatomical sex—that is, for children who are transgender—the feeling of being in the wrong body may intensify in adolescence. The tension between outward appearance—which sets up a host of social expectations regarding behavior—and inner feelings causes confusion and distress, and may make interactions with others difficult. Moreover, transgender adolescents often face severe discrimination and physical danger. As a result, they may experience anxiety, depression, or other disorders.

Neuroscientists are beginning to discover some of the biological factors responsible for transsexuality. To begin with, they have found that anatomical sex and gender identity are determined separately, at different times in the course of development. Sexual differentiation of the genitals, which determines anatomical sex, takes places in the first two months of pregnancy, whereas sexual differentiation of the brain, which governs cognitive aspects of behavior such as gender identity, starts during the second half of pregnancy. Hormones may influence each of these two processes independently, which accounts for the feeling of being in the wrong body. In addition, some genes can cause gender identity to diverge from anatomical sex at birth. No studies have found that social environment after birth has an effect on gender identity.

In puberty, the body produces a cascade of hormones. The hypothalamus releases a hormone that causes the pituitary gland, in turn, to release hormones. The hormones released by the pituitary gland prompt the testes or the ovaries to release what are known as sex hormones: testosterone for boys and estrogen for girls. These sex hormones produce the changes in the male and the female body that are associated with puberty. By the time children reach puberty, 75 percent of those who have questioned their gender will identify as the gender assigned at birth. However, those who identify as transgender later, in adolescence, almost always do so permanently.

Since the 1980s, physicians have been able to intervene at the very beginning of puberty to prevent the release of sex hormones. Besides its success in delaying puberty, this treatment is completely reversible. Thus physicians can ease the psychological trauma of puberty for their patients who want to change their sex to match their gender identity. Giving these patients drugs that block puberty buys them time until their bodies and their decision-making capabilities are mature enough to decide whether they want to begin cross-sex hormone treatment. Without puberty-blocking drugs, adolescents will undergo physical changes that are difficult, if not impossible, to reverse.

In 2017 the Endocrine Society recommended that puberty-blocking therapy be started when boys and girls first exhibit the physical changes associated with puberty. Treatment with cross-sex hormones is not completely reversible and therefore should wait until age 16 in most cases.

Although some people question the idea of providing puberty-blocking drugs to adolescents because their side effects are not well understood, many endocrinologists, psychologists, and others who are involved in this area of treatment believe that denying trans-gender adolescents the ability to transition by withholding the drugs is unethical. Failing to treat adolescents is not simply being neutral, they point out; it means exposing them to serious psycho-logical and possibly physical harm.

Some psychologists encourage children to live as the gender they identify with before puberty. This is an increasingly popular choice, but it is controversial. Many other psychologists discourage such social transitioning until the teenage years. Regardless of the approach to children's gender identity, clinicians and families should help children to understand what they are experiencing as they go through the social and physical transition.

Findings from a recently completed study funded by the National Institutes of Health are beginning to clarify when and how best to help adolescents who are seeking to transition from their sex at birth. The project, a collaboration of four academic medical clinics, is the largest study of transgender youth thus far and only the second study to track the psychological effects of delaying puberty; it is also the first study to track the medical impacts of delaying puberty. One group received puberty-blocking hormones at the beginning of adolescence; another, older group received cross-sex hormones. Ultimately, the study aims to understand whether early medical intervention reduces the health disparities that disproportionately affect transgender people throughout their lives.

We know that transgender people of all ages are often denied basic human rights and often subjected to life-threatening violence. Children and adolescents may be bullied mercilessly. They may not be allowed to use the bathroom that aligns with their gender identity or to play the sports they want to play. In 2020 at least six state legislatures considered bills that would restrict transgender minors' access to hormone therapy. Some of the bills would penalize physicians for administering such therapy; others would classify it as child abuse.

Studies of gender identity offer a clear example of neuroscience as a liberating influence in our lives. A sharper focus on the biology of gender identity will give us a much clearer picture of the range of human sexuality, including our own, and make us more understanding and accepting of transgender men and women. It will also enable us to understand transgender children and adolescents and help them transition into adulthood.

In working on the *Brain Series,* as in writing for a general audience, I found that explaining why a particular area of science is interesting requires a different part of the brain than just reciting facts. And I really had to work to get this right. Surprisingly, I also found that trying to explain a scientific topic to an audience of nonscientists ended up making things clearer for me. I realized that in some instances I had failed to attend to or, worse still, had not properly understood some scientific point in the first place.

Thus, working on a new talk or a new topic for a general audience is worthwhile on a number of levels, including the educational experience it provides for the person who prepares the talks. This was certainly true for me in the *Brain Series*—and I had a wonderful time with it.

THE DISORDERED MIND: WHAT UNUSUAL BRAINS TELL US ABOUT OURSELVES

Inspired by the *Brain Series,* and in an attempt to expand the reach of the television programs, I began work in 2012 on a book about brain disorders for a general readership. That book, *The Disordered Mind: What Unusual Brains Tell Us About Ourselves,* was published in 2018 and was named Science Book of the Year in the field of medical biology by the Austrian Ministry of Education, Science, and Research.

Brain disorders result when some part of the brain's circuitry—the network of neurons and the synapses they form—is overactive, inactive, or unable to communicate effectively. The dysfunction may stem from faulty wiring of the brain during development, injury later in life, or age-related changes in synaptic connections between neurons in the elderly. Depending on what regions of the brain are affected, the disorders change the way we experience life—our emotion, cognition, memory, social interaction, creativity, freedom of choice, movement, or, most often, a combination of these aspects of our biological nature.

In *The Disordered Mind* I explore how processes in the brain can become dysfunctional, resulting in devastating diseases such as autism, depression, bipolar disorder, schizophrenia, Alzheimer's disease, Parkinson's disease, addiction, and post-traumatic stress disorder. I describe what scientists and clinicians have learned about these disorders from observing patients and from neuroscientific and genetic research, and I give readers an opportunity to hear people with these disorders describe their lives and their experiences.

The study of brain disorders gives scientists and clinicians invaluable insights into how the mind works when brain circuits are functioning robustly, and this knowledge helps them develop effective treatments when some of those circuits are compromised. Thanks in large part to advances in genetics, brain imaging, and animal models, scientists studying brain disorders have confirmed several general principles of how our brain normally functions. For example, when one neural circuit in the brain is turned off, a different circuit, which was inhibited by the inactivated circuit, may turn on. We see this principle at work in imaging studies showing that the left and right hemispheres of the brain deal with different mental functions and that the two hemispheres inhibit each other. Specifically, damage to the left hemisphere can free up the creative capabilities of the right hemisphere.

Scientists have also uncovered some surprising links between disorders that appear to be unrelated because they are characterized by dramatically different kinds of behavior. As noted earlier, the symptoms of Parkinson's disease and Alzheimer's disease vary widely because the particular proteins affected in each disorder differ, as do the functions for which they are responsible. Similarly, both autism and schizophrenia involve synaptic pruning—the removal of excess synapses and even dendrites on neurons—during development. In autism, not enough dendrites are pruned, whereas in schizophrenia too many are. In another example, three different disorders—autism, schizophrenia, and bipolar disorder—share genetic variants. That is, some of the same genes that create a risk for schizophrenia also create a risk for bipolar disorder, and a different group of genes that creates a risk for schizophrenia also creates a risk for autism spectrum disorders.

Finding common mechanisms of brain disorders not only helps us understand how the brain works, it also helps in devising treatments and preventive strategies. For example, scientists have used deep-brain stimulation of different regions to regulate neural circuits in depression, on the one hand, and to calm the circuits that have degenerated in Parkinson's disease, on the other.

Neuroscientists have studied how our sense of self comes undone when the brain is assaulted by trauma or disease. This approach implies a "normal" set of behaviors, but the line separating "normal" from "abnormal" has been drawn in different places by different societies throughout history. All variations in behavior arise from individual variations in the brain. My own work has found that learning changes the connections between neurons in the brain. This means that, in addition to the differences between our brains at birth, each brain is further differentiated from every other brain because of what we have learned and remember.

In a larger sense, the biological study of mind is more than a scientific inquiry holding great promise for expanding our understanding of the brain and devising new therapies for disorders of the brain. It opens up new avenues into the study of consciousness and creativity and offers the possibility of a new humanism—one that merges the sciences, which are concerned with the natural world, and the humanities, which are concerned with the meaning of human experience.

The new scientific humanism, based largely on biological insights into differences in brain function, will fundamentally change the way we view ourselves. Each of us feels unique, thanks to our consciousness of self, but now we will have biological confirmation of our individuality. That knowledge, in turn, will enrich our understanding of and empathy with people who think differently and make us less likely to stigmatize or reject them.

CHAPTER 4

INTRODUCING BRAIN
SCIENCE TO ART

The experience of writing *In Search of Memory* and my role as co-host of the *Brain Series* gave me the courage to consider writing about a wider range of ideas. I have always been interested in the humanities and the arts, as well as in science, and Denise and I are modest but avid collectors of art. I have been particularly fascinated by how people perceive and respond emotionally to works of art—and I thought a general reader would also be fascinated by these questions. So I brought these interests together in my first book about the brain and art: *The Age of Insight: The Quest to Understand the Unconscious in Art, Mind, and Brain, From Vienna 1900 to the Present,* published in 2012.

The enjoyment I had from writing this book, and its favorable reception, encouraged me to explore the relation of brain science and art further. So in 2016 I published a second book about art, *Reductionism in Art and Brain Science: Bridging the Two Cultures.* In addition, a chapter in *The Disordered Mind* explores in some depth the link between creativity and brain disorders, particularly instances of remarkable artistic and literary creativity in people with schizophrenia and bipolar disorder. We now know that creativity can arise from the same connections among brain, mind, and behavior present in everyone.

THE AGE OF INSIGHT: THE QUEST TO UNDERSTAND THE UNCONSCIOUS IN ART, MIND, AND BRAIN, FROM VIENNA 1900 TO THE PRESENT

The idea for *The Age of Insight* arose in a rather surprising and indirect fashion. In 1994 I received an honorary degree from the University of Vienna School of Medicine and was asked to make comments on behalf of myself and the two other people receiving awards that day. In preparation, I began to read about the history of the Vienna School of Medicine and discovered the extraordinary contributions to modern medicine made by Carl von Rokitansky, a pathologist and dean of the school.

Rokitansky linked clinical signs and symptoms of a disease observed at a patient's bedside with later pathological findings at autopsy, thereby essentially putting the practice of medicine on a scientific footing. He realized that to get to the truth, he had to go below the surface of the body. The same realization later animated the psychoanalytic theories of Sigmund Freud and the work of the Austrian Expressionist artists Gustav Klimt, Oskar Kokoschka, and Egon Schiele. In fact, it became clear to me that Viennese medicine has contributed a great deal not only to medicine as a whole but also to our modern understanding of mind and brain.

Rokitansky interested me for other reasons as well. His associate, Emil Zuckerkandl, was the husband of Berta Zuckerkandl, who ran one of Vienna's most important salons, bringing artists, scientists, and writers together and thus fostering an easy exchange of ideas. It was at Berta's salon that Klimt learned about biology from Emil and began to incorporate some biological ideas and images into his paintings.

Sometime later, I gave a talk at the Practitioners Club in New York City. Members of this club meet six times a year, and they take turns giving a talk on whatever subject they choose. I thought I would speak about my hobby, the Austrian Expressionist artists, whose work on paper Denise and I collect. The talk was very well received, and as I was speaking, I was struck again by the connection between Rokitansky and the influence he had on these artists through the Zuckerkandl salon.

Finally, for a meeting in Vienna of the European Neuroscience Society, I gave a talk that pulled the whole thing together, beginning with Rokitansky, Zuckerkandl, the influence of Zuckerkandl on Klimt, and so on. The young people who attended didn't know about these artists. They enjoyed the talk, and they ran to the great museums of Vienna to see the Expressionist paintings.

After giving these talks, I was inspired in 2008 to write *The Age of Insight*, a book about Vienna 1900, the place and the time where revolutions in medical science, in psychoanalysis, and in the arts gave rise to the modern era. I focused mainly on Freud, the artists Klimt, Kokoschka, and Schiele, and what the new science of mind—the biology of perception, emotion, and empathy—can tell us about the creative process and about our understanding of art.

The modernist drive to find truth beneath surface appearances is evident not only in Freud's theoretical work indicating that much of our mental life, including our perception of art, is unconscious but also in the portraits by the Austrian Expressionists. Their startling depictions of a subject's unconscious, instinctual life were revolutionary in their time and evoked strong emotional responses in the viewer, as they still do today.

The leaders of the Vienna School of Art History concurred in the belief that the artist should strive to go below surface appearances, to convey not only beauty but also new truths. Influenced in part by Freud's work, the School of Art History had attained international renown for its science-based psychology of art, which introduced the idea of the beholder's share—the idea that art is not complete without the participation of the viewer.

Just how does the viewer—the beholder—participate in art? When a person looks at an image in art (or at any other image, for that matter), the image represents not so much an objective reality as the viewer's perceptions, emotions, imagination, expectations, and knowledge of other images recalled from memory. Thus, when we look at a painting, each of us unconsciously re-creates in our mind the image the artist has painted.

How does the brain do that? Every mental process—perceptual, emotional, or motor—relies on distinct groups of specialized neural circuits located in an orderly, hierarchical arrangement in specific regions of the brain. Visual perception is enormously complex, but simply put, vision begins with a pattern of light that is reflected by an object onto the retina of the eye. That image is first deconstructed into electrical signals that describe lines and contours, creating a boundary around the object and separating it from the background. These initial steps in visual processing are known as bottom-up processing.

As signals move into and through the brain, they are recoded, reconstructed, and elaborated into the image we perceive. This is done through top-down processing, signals sent to lower cognitive centers of the brain by higher cognitive centers. Top-down processing compares incoming visual information with our prior knowledge and experiences and confers significance on the image. Our ability to find meaning in what we see depends entirely on top-down processing—and top-down processing relies on memory.

We now know that one of the reasons the art of the Austrian Expressionists appeals to us so strongly is that our brain is hardwired to respond to exaggerated depictions of faces, bodies, and bodily movement and to understand other people's minds and emotions. The Expressionists' intuitive understanding and careful study of the emotive nature of faces, bodies, and hands allowed them to convey more dramatic psychological portraits than anyone before them had. Moreover, they had mastered the perceptual principles that our eyes and brain use to construct the world around us. With these new insights into the emotional and perceptual apparatuses of the mind, the Austrian Expressionists, in parallel with Freud, knew how to enter another person's mind, to understand its nature, mood, and emotion, and to articulate that understanding to the viewer.

The brain is a creativity machine. It searches for patterns amid chaos and ambiguity, and it uses inferences and guesses to construct models of the complex reality around us. This search for order and pattern is at the heart of the artistic and the scientific enterprise alike.

Today, we are in a position to address directly what neuroscientists can learn from the experiments of artists and what artists and beholders can learn from neuroscientists about artistic creativity, ambiguity, and the perceptual and emotional response of the viewer to art. The role of creativity in the artist and in the beholder is now being actively explored by psychologists and art historians, using the tools of cognitive psychology, and by neuroscientists, using brain imaging. For neuroscientists, the study of creativity ranks with the study of consciousness as being on the edge of the unknown.

Writing *The Age of Insight* gave me enormous pleasure, and that pleasure was enhanced when in 2013 it won the Bruno Kreisky Prize, Austria's highest literary award. Its publication also came at a time when my relationship to Vienna was changing, which I describe below.

In 2015, when the movie *Woman in Gold* came out, sales of my book, which had been respectable, ticked up. The movie was based on the true story of Maria Altmann, whose aunt, Adele Bloch-Bauer, was the subject of Klimt's famous painting of the woman in gold. The painting was stolen by the Nazis, and after many years of struggle with the Austrian government by Maria and her lawyer, E. Randol Schoenberg, it was finally returned to its rightful owners.

After *The Age of Insight* was published, I was invited by the noted art historian Emily Braun to contribute an essay to *Cubism: The Leonard A. Lauder Collection,* a volume published by the Metropolitan Museum of Art on the occasion of its exhibition of Lauder's collection.[1] The exhibition celebrated Lauder's promised gift of the collection to the museum, which described the artworks as "the most significant private holding of Cubist art in the world today." My essay, "The Cubist Challenge to the Beholder's Share," expanded on ideas first introduced in *The Age of Insight*

FRONTISPIECE Eric Kandel receiving the Nobel Prize in Physiology or Medicine, Stockholm, Sweden, December 10, 2000 © The Nobel Foundation, photo: Hans Mehlin

FIGURE INT.1 Denise and me on our wedding day, June 10, 1956.

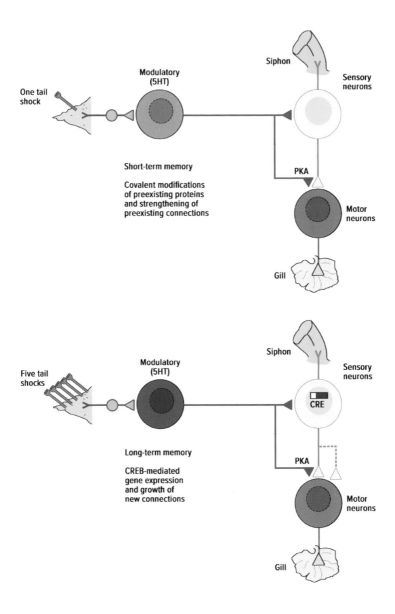

FIGURE 2.1 Molecular mechanisms of memory. Short-term memory involves strengthening existing connections between neurons (above); long-term memory requires the growth of new connections (below). Artist: Sarah Mack

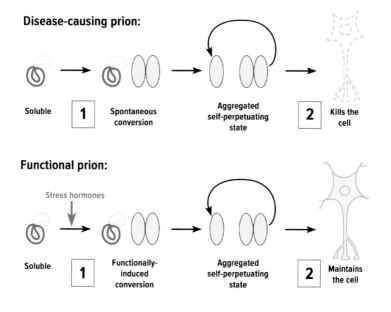

FIGURE 2.2 Comparison of disease-causing and functional prion states. Disease-causing prions undergo a spontaneous conversion to the prion state (above). Stress hormones released in response to an electric stimulus in learning tasks convert CPEB to the prion-like state (below). Artist: Sarah Mack

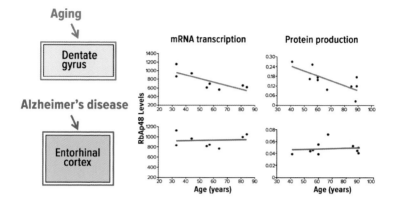

Elias Pavlopoulos et al. 2013

FIGURE 2.3 Gene expression and memory loss. Expression of the RbAp48 gene as measured by mRNA transcription (center) and protein production (right) decreases linearly in the aging human dentate gyrus (green) but not in the entorhinal cortex (red) or other regions of the hippocampus involved in Alzheimer's disease. Artist: Sarah Mack

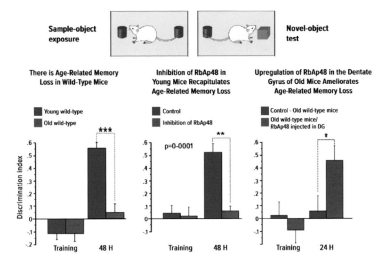

FIGURE 2.4 RbAp48 protein and memory loss. Inhibiting expression of the RbAp48 gene, and thus the amount of RbAp48 protein, in the dentate gyrus of young mice causes memory deficits characteristic of old mice, as indicated in spatial object recognition tasks. Increasing the amount of protein in the dentate gyrus of old wild-type (non-genetically modified) mice ameliorates age-related memory loss. Artist: Sarah Mack

FIGURE 4.1 Arnold Schoenberg, *The Red Gaze,* 1910.

A B C

FIGURE 6.1 (A) Richard Axel; (B) Thomas Jessell; (C) Rui Costa. A and B courtesy of Jill LeVine, C courtesy of John Abbott

FIGURE 6.2 Jerome L. Greene Science Center, Columbia University Manhattanville Campus. Designed by Renzo Piano Building Workshop, with Davis Brody Bond, Executive Architect. Photo: ©Frank Oudeman/ Columbia University

and elaborated upon in my next book, *Reductionism in Art and Brain Science.*

REDUCTIONISM IN ART AND BRAIN SCIENCE: BRIDGING THE TWO CULTURES

My second book on art and neuroscience focuses on the difference between a viewer's responses to abstract art and to figurative art. It, too, was initially inspired by a talk that I gave, as well as by some brain-imaging studies I was working on in collaboration with Daphna Shohamy, a psychologist at Columbia.

In 2014 I was invited by Frieder Burda to speak at the tenth anniversary of his museum, in Baden-Baden, Germany. The museum has an extraordinary collection of abstract art, so I thought I would carry my discussion of art and science one step further and focus on the people who started off as figurative artists and then became abstract artists. Gerhard Richter, a German artist, was one of these, as was the American artist Mark Rothko; both are well represented in the Burda Museum collection. I ultimately ended up expanding on this idea, and in doing so I realized that many of the painters who converted from figurative art to abstract art belonged to the New York School of the 1940s and 1950s. I therefore decided to build my next book mainly around those artists and their move from figurative art to Abstract Expressionism—that is, their move toward reductionsim.

Like most scientists, I have applied the principle of reductionism to my work, in my case by using the simple sea snail, *Aplysia,* to study learning and memory. However, it became clear to me from studying the work of abstract artists that reductionism is not limited to scientists. Abstract artists such as Vassily Kandinsky, Piet Mondrian, and Kasimir Malevich are radical reductionists.

Moreover, artists sometimes use methodologies similar to those used by scientists. The painters of the New York School are a prime example. They often took an investigative, experimental approach to their work. Their attempts to solve the technical problems associated with increased abstraction required them to explore the nature of visual representation.

Brain scientists and cognitive psychologists have studied visual representation at great length and now have a good understanding of how it works. To begin with, our eyes are not a camera. While they do collect the information we need to act, our eyes do not present the brain with a finished product. Instead, our brain actively extracts information about the three-dimensional organization of the world from the two-dimensional images that are projected onto the retina of the eye. The brain recodes that information and, based on its rules of organization and on our prior experience, reconstructs and elaborates on the image.

It is astonishing—indeed, almost magical—that our brain can perceive an object based on such incomplete information. It can even perceive that object as being the same under strikingly different conditions of lighting and context. Moreover, each of us is able to create a rich, meaningful image of the external world that is remarkably similar to the image seen by others. It is in the construction of these internal representations of the visual world that we see the brain's creative process at work.

But abstract art does not rely on the brain's visual system to re-create an image. Instead, it dares our brain to interpret an image that is fundamentally different from the kind of images the brain has evolved to reconstruct. To present a subjective vision and state of mind rather than a naturalistic representation of the outside world, the abstract artist dismantles many of the building blocks of the brain's visual processing system. For instance, the artist might eliminate links between line segments, or recognizable contours and objects, or perspective. Sometimes the artist reduces an image to its essence, presenting form, line, color, or light in isolation. The isolated element stimulates aspects of our imagination in ways that a complex image might not and can thereby produce a strong emotional, even spiritual, impact. Thus the reason abstract art poses such an enormous challenge to the beholder is that it teaches us to look at art—and, in a sense, at the world—in a new way.

In abstract painting, elements in the painting are intended as references or clues to how we conceptualize objects: the artist attempts to create conditions that enable us to complete the picture based on our own unique experience. Put another way, abstract art relies on the assumption that an impression, a sensory stimulation of the retina, sparks a creative process of associative recall in the beholder. Evidence from brain studies supports this idea of associative recall. In fact, the pleasure that many viewers derive from abstract art seems to stem in good part from the experience of associating something never quite seen before with familiar images. I believe that the beholder's creative reconstruction of the image is inherently pleasurable.

Thus, artists and scientists alike are reductionists, but they have different ways of knowing and making sense of the world. Scientists make models of elementary features of the world that can be tested and reformulated. These tests rely on removing the subjective biases of the observer and using objective measurements or evaluations. Artists form models of the world, but theirs are subjective impressions of the ambiguous reality they encounter in their everyday lives. The brain's ability to construct models of the world makes possible both the artist's creation and the beholder's re-creation of a work of art. Both derive from the intrinsically creative workings of the brain.

Art is best understood as a distillation of experience. As such, it complements and enriches the science of mind. Neither approach alone is sufficient to fully understand the dynamic of human experience. What we require is a third way, a set of explanatory bridges across the chasm between art and science.

Modern brain science can help bridge that chasm. Although brain science is at a very early stage of exploring the biological underpinnings of our response to art, we do have some clues as to why abstract art can be successful in eliciting an active, creative, and enriching response in the beholder. Moreover, brain science has revealed the cellular and molecular underpinnings of learning and memory. Those same biological mechanisms are central to our response to a work of art.

When I am asked if applying the reductionist approach of brain science to the study of art takes the magic out of art, I reply, "Not in the slightest." Nor does appreciating the reductionist methods used by artists in any way diminish the richness or complexity of our response to art. In fact, some artists have used a reductionist approach to explore the foundations of creativity. So yes, brain science enhances our understanding—and our enjoyment—of art.

Not long after *The Age of Insight* came out, and before *Reductionism in Art and Brain Science* was published, Randy Schoenberg, the lawyer who recovered Klimt's painting of Adele Bloch-Bauer, sent me an email saying, "Nu, where's my grandfather?" His grandfather was the great composer Arnold Schoenberg. I answered saying, "Your grandfather was a great artist," which most people don't realize, "but he didn't fit into this book." I did come back to Randy's grandfather in *Reductionism in Art and Brain Science,* because Arnold Schoenberg is part of an interesting and complicated story that I describe in that book.

Schoenberg was not earning much money as a composer. He therefore began to entertain the idea that he would become a painter, because he believed that painters earned money very easily. Schoenberg was very impressed with a young artist named Richard Gerstl, who was moving in an Expressionist direction before Oskar Kokoschka, and before Schiele began to paint himself in the nude. Although Gerstl was very young, in his early 20s, he was really quite an extraordinary painter, so Schoenberg asked Gerstl to give him some lessons. Schoenberg then invited the artist to come with him to his summer residence and spend the season there.

Gerstl went for the summer and proceeded to have an affair with Madam Schoenberg. Schoenberg found out about it and gave his wife a choice: him or me. She chose Schoenberg, and Gerstl—age 26—killed himself. Before he did so, however, he painted a portrait of her with her lips sealed and her eyes closed, because she had been told never to look at Gerstl again.

After the death of Gerstl, Schoenberg went on to paint for about three or four years, and by 1910 his work had become quite remarkably abstract (fig. 4.1). Around this time, Kandinsky was trying to leave figuration but having difficulty doing so. In January 1911 he went to a New Year's concert and heard Schoenberg's atonal music—at which point Kandinsky said to himself, *If Schoenberg can leave tonality, I can leave figuration.* He began his abstract work in April 1911, not knowing that Schoenberg was also an artist and that, as an artist, Schoenberg had beaten him to it.

FIGURE 4.1 Arnold Schoenberg, *The Red Gaze,* 1910.

CHAPTER 5

RETURN TO AUSTRIA

N ot long after winning the Nobel Prize, I received a telephone call from Thomas Klestil, then the president of Austria, saying, "Isn't it wonderful that we have another Austrian Nobel Prize!" I politely informed him that I left Austria at age 9, did all my work in the United States, and considered myself a Jewish American scientist.

Later, he wrote me a letter asking, "What can we do to honor you?" And I said, "I have more honors than I deserve. I would like to see a symposium at the University of Vienna on the response of Austria to National Socialism." He replied that he would be delighted to have such a symposium. Fritz Stern, a noted historian of modern German culture and politics from Columbia University, helped me choose participants. Two friends, Anton Zeilinger, an outstanding quantum physicist from the University of Vienna, and Friedrich Stadler, of the Institute Vienna Circle, helped to conceptualize and organize the event.

The symposium, entitled "Austria and National Socialism: Implications for Scholarship in Science and the Humanities," was held in June 2003. Its goal, as described by Stadler, was "to convey to the general public and to the scientific community, in particular the younger generation of scholars and students, the disastrous effects of Nazi rule in Austria—expropriation, expulsion, and the Holocaust—on the entire field of education and research."[1] In addition, we intended the symposium to compare Austria's response to National Socialism with that of Germany, France, and Switzerland and to highlight Austria's delayed recognition of its key role in the Nazi efforts to annihilate the Jews. Zeilinger and others proposed a scholarship program for young Jewish scientists and visiting professorships for researchers expelled from Austria by the Nazis.

Altogether, the event was really quite wonderful and well attended. In 1996 the chancellor of Austria had made a formal apology about the role that Austria played in the Nazi period, and at this symposium people spoke honestly for the first time about how horrible Austria's interactions with the country's Jews were.

The symposium gave me the opportunity to establish contact with the Jewish community in Vienna and to discuss with them how they viewed their life and what they thought the symposium had accomplished. The Jewish community agreed that the symposium had helped young Viennese academics recognize Austria's enthusiastic collaboration with the German Nazis, and, further, had focused international attention—through newspapers, television, radio, and magazines—on Austria's active role during the Hitler era and the Holocaust.

In September 2004 Denise and I went back to Vienna to celebrate the publication of a book based on the symposium. Klestil had died earlier that year, and the new president, Heinz Fischer, asked us to dinner at the Sacher Hotel. So he and his wife, Denise and I, and Zeilinger and his wife, the six of us, had dinner at the hotel, with no guards or any other protection.

As we were having the second glass of wine, Denise asked Heinz Fischer whether the president of Austria is just an honorific office or has real power. Heinz responded powerfully with: "Of course I can declare war immediately, but I'm not going to do that." After our laughter subsided, Denise took him through his political career, which was really quite fascinating. His wife is part Jewish and was in hiding in Sweden during the war. Heinz was also in Sweden during the war. He'd been in Austrian politics for a very long time. He had a strong education in science, and he'd done a great deal for science in Austria. In fact, at one point he was minister of science.

So we became friends, and now almost every time we're in Vienna he invites us for a cup of tea. It's just wonderful to think that you're kicked out of Austria one day, and now you have tea with the president! In fact, I went on to help change the name of the street that the university is on from Karl Lueger Platz to University Platz. Lueger, the mayor of Vienna from 1897 to 1910, was an outspoken anti-Semite.

My relationship with Austria is becoming more comfortable, although it's got a way to go. I've been helping two of the neuroscientific organizations in Vienna by reviewing their scientific program. In return, I have received an Austrian Medal of Honor for Science and Art, as well as a Decoration of Honor for Service to Austria, and in 2009 I was named an honorary citizen of Vienna. In 2013 I wrote an essay on the productive interaction of Christians and Jews for another symposium, this one on "The Long Shadow of Anti-Semitism at the University of Vienna Dating Back to 1870," a name that reflects the transparency that has come to characterize Austria's new attitude toward its past. In 2018 I was awarded an honorary doctorate by the Medical University of Vienna; also in that year the city of Vienna placed a commemorative plaque on the apartment where my family and I lived during my youth.

Denise and I usually go to Vienna once a year. In addition, we took advantage of the funds from the Nobel Prize and bought ourselves an apartment in Paris, which we enjoy enormously and to which we have gone almost every Christmas. It hardly pays back what Denise has given me, but it's really been a great joy to us both.

CHAPTER 6

COLUMBIA UNIVERSITY AND THE SCIENCE OF MIND, BRAIN, BEHAVIOR

C olumbia University was a natural choice for the Howard Hughes Medical Institute when it decided in 1984 to focus part of its biomedical research funding on neuroscience. Columbia was one of the first universities to bring together researchers from different fields to study behavior at the cellular, molecular, and systems levels. With time, neural science at Columbia succeeded in merging the fields of cell biology, physiology, and development of the nervous system with molecular biology, including molecular genetics. Recently, Columbia has unified this cell and molecular approach to the brain by including cognitive psychology and theoretical neural science systems.

In 2004 a Kavli Institute for Brain Science was established at the Columbia University Medical Center with a $7.5 million gift from the Kavli Foundation. In that same year, the university announced the formation of a Mind Brain Behavior Initiative that would pull together all of Columbia's research on the brain, cognition, and behavior. In early 2006 the late philanthropist Dawn M. Greene and the Jerome L. Greene Foundation gave Columbia $250 million, the largest gift in the university's history, to go forward with building plans for the Jerome L. Greene Science Center, which would house the faculty and students participating in the new Mind Brain Behavior Initiative. In 2012, with a $200 million endowment from businessman and philanthropist Mortimer B. Zuckerman, the initiative became the Mortimer B. Zuckerman Mind Brain Behavior Institute.

THE KAVLI INSTITUTE FOR BRAIN SCIENCE

Work at the Kavli Institute focuses on developing more powerful experimental and computational tools for studying the brain's complex neural circuits. Such tools enable us to move from the study of individual nerve cells to the study of the complex neural systems that underlie cognitive function and behavior. The institute uses advanced imaging technology to observe individual neurons, synapses, and neural circuits as they develop and interact, and as they are modified by learning and memory.

Research has shown that information encoded by the brain's neural circuits holds the key to understanding attention, perception, emotion, action, and memory. In turn, understanding how the brain works will provide us insights that may one day lead to treatments for a wide range of disorders of brain circuitry, such as epilepsy, Alzheimer's disease, Parkinson's disease, amyotrophic lateral sclerosis (ALS), schizophrenia, and autism.

With members from a wide range of disciplines, the Kavli Institute unites the core sciences of chemistry, biology, and physics with the disciplines of philosophy, psychology, engineering, business, law, and economics to address questions of how the brain works and how it affects the way we think, perceive, and interact with the world. In addition to its links within the neuroscience community at Columbia, the institute connects with six other Kavli neuroscience institutes at other universities. When the Kavli Institute at Columbia was established, I was named Kavli Professor in the Department of Neuroscience and director of the Kavli Institute.

THE MORTIMER B. ZUCKERMAN MIND BRAIN BEHAVIOR INSTITUTE

Just as science in the twentieth century was transformed by the discovery of DNA, so science and medicine in the twenty-first century is being transformed by what we are learning about the mind, paving the way to improved brain health and greater understanding of what it means to be human. This research goes beyond the search for solutions to devastating illnesses. The new science of mind attempts to penetrate the mystery of consciousness, including the ultimate mystery: how each person's brain creates the consciousness of a unique self and the sense of free will.

Our work at the Zuckerman Mind Brain Behavior Institute focuses on understanding how the brain works and how it gives rise to our mind and the complexities of behavior, providing deeper insights into our mental functions in both health and disease. A key goal is translating basic research findings into new therapies and potential cures for brain disorders such as Alzheimer's disease, Parkinson's disease, ALS, autism, schizophrenia, mood disorders, memory loss, traumatic brain injury, and stroke. A closely related goal, developing accurate diagnoses and therapies tailored to individual patients, relies on research that explores the connections between genetic variations and brain disorders.

These ambitious undertakings are based on the recognition that every area of study in a university concerns the mind. Thus the Mind Brain Behavior Institute envisions itself as the hub of research on brain science at Columbia, bringing together faculty and students, including some undergraduates, from the neural sciences, social sciences, humanities, arts, economics, law, and engineering to develop new insights into the nature of thinking, acting, and being. As Columbia University President Lee Bollinger said, "This is a remaking of the intellectual life of the University."

Together with Richard Axel and Thomas Jessell, I was honored to be named one of the founding co-directors of the Mind Brain Behavior Institute. Today, Richard and I continue as co-directors, while Rui Costa is director and chief executive officer (fig. 6.1).

In 2007 the Center for Neurobiology and Behavior was replaced by the Department of Neuroscience, which oversaw the Interdisciplinary Doctoral Program in Neurobiology and Behavior, the

FIGURE 6.1 (A) Richard Axel; (B) Thomas Jessell; (C) Rui Costa.
A and B courtesy of Jill LeVine, C courtesy of John Abbott

Center for Theoretical Neuroscience, the Mahoney-Keck Center for Brain & Behavior, the Kavli Institute for Brain Science, and the Grossman Center for the Statistics of the Mind. Then, in 2018, the faculty of the Department of Neuroscience joined other researchers to form the Zuckerman Mind Brain Behavior Institute, fulfilling Columbia's vision of an interdisciplinary axis for research on the science of mind.

THE JEROME L. GREENE SCIENCE CENTER

The Jerome L. Greene Science Center is essential for realizing the extraordinary potential of the Mind Brain Behavior Institute. Designed by the renowned architect Renzo Piano, the Greene Center facilitates the kind of social interaction and exchange of ideas among a diversity of scholars that is essential for new interdisciplinary concepts to arise and thrive. The 450,000-square-foot building, which opened in 2017, houses 56 laboratories, including my own (fig. 6.2). It also connects the university with its surroundings in an entirely new way. I was delighted to be invited by Lee Bollinger, along with my colleagues Richard Axel and Thomas Jessell, to help plan this remarkable enterprise.

The Greene Center is one of several buildings the university is constructing on its new Manhattanville campus, and in all of them Piano's architecture reflects the synergy between Columbia and the residents of West Harlem. In addition, the plan for the Manhattanville campus is the first to have been awarded New York City's LEED-ND Platinum designation for environmentally sustainable urban design and the U.S. Green Building Council's highest certification for sustainable development.

FIGURE 6.2 Jerome L. Greene Science Center, Columbia University Manhattanville Campus. Designed by Renzo Piano Building Workshop, with Davis Brody Bond, Executive Architect.
Photo: ©Frank Oudeman/Columbia University

Piano spent many hours in conversation with researchers before sketching designs for the Greene Center. He learned that we need quiet spaces for concentration and open spaces where people and ideas from various disciplines can come together. In addition, we need shared public spaces for mixing with people from elsewhere in the university and the surrounding community.

Piano's final designs for the center reflect these needs. The building projects a sense of scholarship while radiating openness, transparency, and accessibility to the public. The first floor houses an education lab where schoolchildren, their teachers, and visitors can learn about the mysteries of brain science, as well as a restaurant and a wellness center where people can get free health services. The upper floors are home to the Mind Brain Behavior Institute. The building's glass façade gives scientists inspiring views of the city, while enabling passersby to observe the researchers working in their laboratories. "Its transparency is meant to underscore that the knowledge being generated inside is for the public and will be shared with them," says Piano.

My colleagues and I are deeply grateful to Columbia University and to the Howard Hughes Medical Institute, two great institutions that have created open environments strongly supportive of scholarship and research. By joining forces, Columbia and HHMI have created an intellectual and physical environment that encourages dialogue among the arts, sciences, and humanities, thus sparking creativity and innovation that will expand our understanding of ourselves and our place in the world.

CONCLUSION

The Nobel Prize is usually awarded to people in a mature stage of their career. After they receive the prize, their work gets even more attention than it did before, so there are many opportunities to give lectures, to participate in symposia, and to write reviews. This has been my experience also, but for me the most surprising result of the Nobel Prize was the interest that my work began to have for journalists, writers of popular science, and producers of television programs, people who communicate with the general public.

Working with these communicators encouraged me to look at my own science from a new, broader point of view and allowed me to connect with a very different audience than I had early in my career. Thus, the Nobel Prize was not only a satisfying recognition of my work, it was also an unanticipated means of introducing neuroscience to a much larger audience, one that is eager to learn about brain science and its role in their lives and in social policy.

In the process of reaching out to the public I have, surprisingly, ended up where I began—as a would-be intellectual historian. I have essentially picked up where I left off upon graduating from Harvard in 1952 with a major in history and literature, proving once more that life comes full circle. Moreover, since 1952 the general public's interest in science has grown significantly. I have enjoyed that greatly. I receive a great deal of satisfaction in explaining science to the general public, and I can see myself as doing more of this in the future.

I assure you, *je ne regrette rien*. Indeed, I want to reassure those of you who are in the running: There is life after the Nobel Prize. In fact, a most enjoyable and fulfilling life. And I think that Denise is slowly but surely becoming convinced that I am not yet dead intellectually, at least not completely.

ACKNOWLEDGMENTS

I am grateful to the wonderful team at Columbia University Press, as well as to my agent, Andrew Wylie, for the help they gave me with this book. I also have benefited greatly from the critical insights on this volume from many of my colleagues at Columbia, including Steve Siegelbaum, Scott Small, and Mickey Goldberg. I am again deeply indebted to my editor, Blair Burns Potter, who worked with me on four earlier books and once again brought her critical eye and her insightful editing to this book. I am also very much indebted to the late Sarah Mack for her editorial suggestions and her outstanding development of the scientific images; all of us who worked with her will miss her greatly. Finally, my great thanks to Pauline Henick, who patiently typed the many versions of this book and, with the aid of Christina Doyle, skillfully guided it to completion.

AWARDS

have been honored with many awards over the years, but in September 2015 I received a particularly unexpected honor: a high school in Ahrensburg, Germany, was named after me. I was told that I was chosen not only because I had won the Nobel Prize for my work in science but also because I was a refugee born to Austrian Jewish parents who had fled to the United States in 1939. This was seen as a wonderful fit for the school's stated goals of high achievement, social responsibility, and diversity.

As Gerd Burmeister, principal of the high school, explained it, "This is one of the leading neuroscientists of our time, and he is the namesake of our school. Why? We wanted someone who has set an example through community involvement, can identify with both students and teachers, and whose name reflects the diversity of our school." Denise and I attended the dedication and entertained the students with a little dancing to the music of the school band.

The other awards I have received fall into two categories: scientific awards and literary awards.

SCIENTIFIC AWARDS

1959   Henry L. Moses Research Award, Montefiore Hospital

1977   Lester N. Hofheimer Prize for Research
        (awarded by the American Psychiatric Association)

        Lucy G. Moses Prize for Research in Basic Neurology
        (awarded by Columbia University)

1978   The Dean's Award for Outstanding Contributions to
        Teaching
        (awarded by Columbia University)

1979   Solomon A. Berson Medical Alumni Achievement
        Award in Basic Science
        (awarded by New York University)

1981   Karl Spencer Lashley Prize in Neurobiology
        (awarded by the American Philosophical Society)

1982   The Dickson Prize in Biology and Medicine
        (awarded by the University of Pittsburgh)

        The New York Academy of Sciences Award in
        Biological and Medical Sciences

1983   Albert Lasker Basic Medical Research Award
        (shared with Vernon B. Mountcastle)

1984   Lewis S. Rosenstiel Award for Distinguished Work in
        Basic Medical Research
        (awarded by Brandeis University and shared with
        Daniel Koshland)

Howard Crosby Warren Medal
(awarded by the Society of Experimental Psychologists)

1985   Association of American Medical Colleges Award for
Distinguished Research in the Biomedical Sciences

1986   Special Presidential Commendation of the American
Psychiatric Association

1987   Gairdner International Award for Outstanding
Achievement in Medical Science
(awarded by the Gairdner Foundation, Canada)

1988   National Medal of Science Gold Medal for Scientific
Merit
(awarded by the Fondazione Giovanni Lorenzini,
Milan, Italy)

National Academy of Sciences for Scientific Reviewing

1989   Distinguished Service Award of the American
Psychiatric Association

Award in Basic Science, American College of Physicians

Robert J. and Claire Pasarow Foundation Award in
Neuroscience

1990   Diploma Internacional Cajal
(Instituto Cajal: Consejo Superior de Investigaciones
Científicas, Madrid)

1991   Bristol-Myers Squibb Award for Distinguished
Achievement in Neuroscience Research
(with T.V.M. Bliss)

1992 John P. McGovern Lectureship Award in Behavioral
      Neuroscience
      (awarded by the American Association for the
      Advancement of Science)

      Warren Triennial Prize
      (awarded by Massachusetts General Hospital)

      Jean-Louis Signoret's Prize on Memory
      (awarded by the Fondation Ipsen, Paris)

1993 Harvey Prize
      (awarded by the Technion, Israel Institute of
      Technology, Haifa, Israel)

      F. O. Schmitt Medal and Prize in Neuroscience
      (awarded by Rockefeller University)

1994 Isaac Ray Decade of Excellence Award
      (awarded by Brown University)

      Mayor's Award for Excellence in Science and
      Technology

      Honorary Degree University of Vienna School of
      Medicine

1995 Stevens Triennial Prize
      (awarded by Columbia University)

1996 New York Academy of Medicine Award

1997 Gerard Prize for Outstanding Achievement in
      Neuroscience
      (awarded by the Society of Neuroscience)

Charles A. Dana Award for Pioneering Achievement in Health
(shared with P. Greengard)

1999   Wolf Prize in Biology and Medicine, Israel

2000   Dr. A. H. Heineken Prize for Medicine
(awarded by the Royal Netherlands Academy of Arts and Sciences, Amsterdam, Netherlands)

Nobel Prize in Physiology or Medicine
(shared with Paul Greengard and Arvid Carlsson)

Distinguished Investigator Award, National Alliance for Research on Schizophrenia and Affective Disorders

2001   Lifetime Achievement Award, YIVO Institute for Jewish Research

Annual Achievement Award, Parkinson Foundation

Office of the Mayor, City of New York, Proclamation May 11, 2001 as "Eric Kandel Day"

Distinguished Service Award, American College of Neuropsychopharmacology

NAMI Pioneer in Science Award

2002   Julius Axelrod Neuroscience Award NARSAD
(shared with Arvid Carlsson and Paul Greengard)

Centenary Medal, Royal Society of Canada

Honorary Member, Alumni Association, Columbia College of Physicians & Surgeons

2003   Honorary Fellow, Distinguished Service in Psychiatry, American College of Psychiatrists

Paul Hoch Award

Distinguished Alumnus Award, New York University Alumni Association

Sven Berggrens Pris, Lund, Sweden

Pupin Medal for Service to the Nation, Columbia University

Salmon Award, New York Academy of Medicine

Benjamin Franklin Creativity Laureate Award, Smithsonian Associates and Creativity Foundation

David Dean Brockman Lectureship Award, American College of Psychoanalysts

2005   Austrian Medal of Honour for Science and Art (presented by the President of the Republic of Austria)

2006   Biotechnology Achievement Award, NYU School of Medicine

Benjamin Franklin Medal for Distinguished Achievement in the Sciences, American Philosophical Society

McKnight Foundation Recognition Award, McKnight Conference for Neuroscience

Louise T. Blouin Foundation Global Creative Leadership Award

2007   Cosmos Club McGovern Award in Science,
        Washington, D.C.

2009   Ellis Island Family Heritage Award

        Honorary Citizenship of the City of Vienna, Austria

        Honorary Award from the Viktor Frankl Foundation
        of the City of Vienna for the Advancement of Meaning-
        Oriented Humanistic Psychotherapy

        Ulysses Medal, University College Dublin, Ireland

        Inauguration of the Eric Kandel Young Scientist Prize,
        Frankfurt, Germany

        An award established by the Hertie Foundation to
        recognize and support the outstanding work of young
        European neuroscientists

        Gold Medal for Distinguished Service to Humanity,
        The National Institute of Social Sciences

2010   Alexander Award in Psychiatry, Baylor College of
        Medicine

        Catcher in the Rye Humanitarian Award, American
        Academy of Child & Adolescent Psychiatry

        Eastman Medal, University of Rochester

2011   National Leadership Award in Science and Education,
        Merage Foundation for the American Dream

2012   Child Mind Institute Distinguished Scientist
        Award

Columbia Lamplighter Award, Chabad at Columbia University

Adolf Meyer Award, American Psychiatric Association

NEPA Distinguished Contribution Award

David Mahoney Prize, Harvard Mahoney Neuroscience Institute

2014 American Association of Chairs of Departments of Psychiatry President's Award

Productive Lives Award, Brain & Behavior Research Foundation

2015 Leo Baeck Medal, Leo Baeck Institute

2016 The Great Medal of Honor of the Vienna Chamber of Physicians

2018 Commemorative plaque placed in the apartment house on Severingasse 8 in the 9th District where Eric Kandel and his family lived

Honorary Doctorate, Medical University of Vienna

LITERARY AWARDS

1982 American Medical Writers Association Book Award (for *Principles of Neural Science*)

1988 J. Murray Luck Award for Scientific Reviewing (awarded by the National Academy of Sciences)

2007   National Academy of Sciences Communication Award: Best Book Award, Irvine, CA (for *In Search of Memory*)

Los Angeles Times 2006 Book Prize for Science and Technology, Los Angeles, CA (for *In Search of Memory*)

NAMI Ken Book Award, NY (for *In Search of Memory*)

2009   Society for Neuroscience Award for Education in Neuroscience (for *Principles of Neural Science IV*) awarded to Eric Kandel, James Schwartz, and Thomas Jessell at the Annual Meeting for the Society for Neuroscience, Chicago, IL

2012   Lifetime Achievement Award, Jewish Book Council

2013   Bruno-Kreisky-Preis für das Politische Buch (Bruno Kreisky Prize for Political Book), Karl Renner Institut, Vienna, Austria (for *The Age of Insight: A Quest to Understand the Unconscious in Art, Mind, and Brain, from Vienna 1900 to the Present*)

2018   Science Book of the Year (Category: Medicine/Biology), Austrian Ministry of Education, Science, and Research (for the German translation of *The Disordered Mind: What Unusual Brains Tell Us About Ourselves.*)

NOTES

1. MOVING TO COLUMBIA AND THE HOWARD HUGHES MEDICAL INSTITUTE

1. E. R. Kandel, "A New Intellectual Framework for Psychiatry," *American Journal of Psychiatry* 155 (1998): 457–69. DOI: 10.1176 /ajp.155.4.457.

2. FURTHER ADVANCES IN SCIENCE

1. K.C. Martin, A. Casadio, H. Zhu, E. Yaping, J. C. Rose, M. Chen, C. H. Bailey, and E. R. Kandel, "Synapse-specific, Long-Term Facilitation of Aplysia Sensory to Motor Synapses: A Function for Local Protein Synthesis in Memory Storage," *Cell* 91 (1997): 927–38. DOI: 10.1016/S0092-8674(00)80484-5.

2. L.E. Hake and J. D. Richter, "CPEB Is a Specificity Factor That Mediates Cytoplasmic Polyadenylation During Xenopus Oocyte Maturation," *Cell* 79 (1994): 617–27.

3. K. Si, M. Giustetto, A. Etkin, R. Hsu, A. M. Janisiewicz, M. C. Miniaci, J. H. Kim, H. Zhu, and E. R. Kandel, "A Neuronal Isoform of CPEB Regulates Local Protein Synthesis and Stabilizes Synapse-Specific Long-Term Facilitation in Aplysia," *Cell* 115 (2003a): 893–904. DOI: 10.1016/S0092-8674(03)01021-3. K. Si,

S. Lindquist, and E. R, Kandel, "A Neuronal Isoform of the Aplysia CPEB Has Prion-Like Properties," *Cell* 115 (2003b): 879–91. DOI: 10.1016/S0092-8674(03)01020-1.

4. K. Keleman, S. Krüttner, M. Alenius, and B. J. Dickson, "Function of the Drosophila CPEB Protein Orb2 in Long-Term Courtship Memory," *Nature Neuroscience* 10, no. 12 (2007): 1587–93. A. Majumdar, W. Colón-Cesario, E. White-Grindley, H. Jiang, F. Ren, M. R. Khan, L. Li, E. M. Choi, K. Kannan, F. Guo, J. Unruh, B. Slaughter, and K. Si, "Critical Role of Amyloid-Like Oligomers of Drosophila Orb2 in the Persistence of Memory," *Cell* 148, no. 3 (2012): 515–29. P. Rajasethupathy, I. Antonov, R. Sheridan, S. Frey, C. Sander, T. Tuschl, and E. R. Kandel, "A Role for Neuronal piRNAs in the Epigenetic Control of Memory-Related Synaptic Plasticity," *Cell* 149, no. 3 (2012): 693–707. E. Pavlopoulos, S. Jones, S. Kosmidis, M. Close, C. Kim, O. Kovalerchik, S.A. Small, and E. R. Kandel, "Molecular Mechanism for Age-Related Memory Loss: The Histone-Binding Protein RbAp48," *Science Translational Medicine* 5, no. 200 (2013): 200ra115. DOI: 10.1126/scitranslmed.3006373. L. Fioriti, C. Myers, Y. Huang, X. Li, J. S. Stephan, P. Trifilieff, L. Colnaghi, S. Kosmidis, B. Drisaldi, E. Pavlopoulos, and E. R. Kandel, "The Persistence of Hippocampal-Based Memory Requires Protein Synthesis Mediated by the Prion-Like Protein CPEB3," *Neuron* 86, no. 6 (2015): 1433–48.

5. K. Si, Y. B. Choi, E. White-Grindley, A. Majumdar, and E. R. Kandel, "Aplysia CPEB Can Form Prion-Like Multimers in Sensory Neurons That Contribute to Long-Term Facilitation," *Cell* 140 (2010): 421–35. DOI: 10.1016/j.cell.2010.01.008.

6. D. Kandel, "Stages in Adolescent Involvement in Drug Use," *Science* 190, no. 4217 (1975): 912–14. DOI: 10.1126/science.1188374.

7. D. Kandel, "Does Marijuana Use Cause the Use of Other Drugs?" *Journal of the American Medical Association* 289, no. 4 (2003): 482–83. DOI: 10.1001/jama.289.4.482.

8. E.A. Griffin Jr., P. A. Melas, R. Zhou, Y. Li, P. Mercado, K. A. Kempadoo, S. Stephenson, L. Colnaghi, K. Taylor, M. C. Hu, E. R. Kandel, and D. B. Kandel, "Prior Alcohol Use Enhances Vulnerability to Compulsive Cocaine Self-Administration by Promoting Degradation of HDAC4 and HDAC5," *Science Advances* 3, no. 11 (2017): e1701682. DOI: 10.1126/sciadv .1701682.

9. A. Levine, Y. Huang, B. Drisaldi, E. A. Griffin, D. D. Pollak, S. Xu, D. Yin, C. Schaffran, D. B. Kandel, and E. R. Kandel, "Molecular Mechanism for a Gateway Drug: Epigenetic Changes Initiated by Nicotine Prime Gene Expression by Cocaine," *Science Translational Medicine* 3, no. 107 (2011): 107ra109. DOI: 10.1126/scitranslmed.3003062.

10. D. Kandel and E. R. Kandel, "A Molecular Basis for Nicotine as a Gateway Drug," *New England Journal of Medicine* 371 (2014): 932–43. DOI: 10.1056/NEJMsa1405092.

11. C. Kellendonk, E. H. Simpson, H. J. Polan, G. Malleret, S. Vronskaya, V. Winiger, H. Moore, and E. R. Kandel, "Transient and Selective Overexpression of Dopamine D2 Receptors in the Striatum Causes Persistent Abnormalities in Prefrontal Cortex Functioning," *Neuron* 49 (2006): 603–15. DOI: 10.1016/j.neuron.2006.01.023.

12. Pavlopoulos et al., "Molecular Mechanism for Age-Related Memory Loss," 200ra115.

3. ADVENTURES IN THE PUBLIC
UNDERSTANDING OF SCIENCE

1. E. R. Kandel and W. A. Spencer, "Cellular Neurophysiological Approaches in the Study of Learning," *Physiological Review* 48, no. 1 (1968): 65–134. DOI: 10.1152/physrev.1968.48.1.65.

2. Charlie Rose, *The Brain Series,* 2009–2017, accessed April 22, 2021, https://charlierose.com/collections/3.

4. INTRODUCING BRAIN SCIENCE TO ART

1. E. Braun and R. Rabinow, *Cubism: The Leonard A. Lauder Collection* (New York: Metropolitan Museum of Art, 2014).

5. RETURN TO AUSTRIA

1. https://www.austria.org/jn-july-2003.

REFERENCES

Braun, E., and R. Rabinow. *Cubism: The Leonard A. Lauder Collection*. New York: Metropolitan Museum of Art, 2014.

Charlie Rose. *The Brain Series*. 2009–2017. Accessed April 22, 2021. https://charlierose.com/collections/3.

Fioriti, L., C. Myers, Y. Huang, X. Li, J. S. Stephan, P. Trifilieff, L. Colnaghi, S. Kosmidis, B. Drisaldi, E. Pavlopoulos, and E. R. Kandel. "The Persistence of Hippocampal-Based Memory Requires Protein Synthesis Mediated by the Prion-Like Protein CPEB3." *Neuron* 86, no. 6 (2015): 1433–48.

Griffin, E. A. Jr., P. A. Melas, R. Zhou, Y. Li, P. Mercado, K. A. Kempadoo, S. Stephenson, L. Colnaghi, K. Taylor, M. C. Hu, E. R. Kandel, and D. B. Kandel. "Prior Alcohol Use Enhances Vulnerability to Compulsive Cocaine Self-Administration by Promoting Degradation of HDAC4 and HDAC5." *Science Advances* 3, no. 11 (2017): e1701682. DOI: 10.1126/sciadv.1701682.

Hake, L. E., and J. D. Richter. "CPEB Is a Specificity Factor That Mediates Cytoplasmic Polyadenylation During Xenopus Oocyte Maturation." *Cell* 79 (1994): 617–27.

Kandel, D. "Does Marijuana Use Cause the Use of Other Drugs?" *Journal of the American Medical Association* 289, no. 4 (2003): 482–83. DOI: 10.1001/jama.289.4.482.

——. "Stages in Adolescent Involvement in Drug Use." *Science* 190, no. 4217 (1975): 912–14. DOI: 10.1126/science.1188374.

—— and E. R. Kandel. "A Molecular Basis for Nicotine As a Gateway Drug." *New England Journal of Medicine* 371 (2014): 932–43. DOI: 10.1056/NEJMsa1405092.

Kandel, E. R. "A New Intellectual Framework for Psychiatry." *American Journal of Psychiatry* 155 (1998): 457–69. DOI: 10.1176/ajp.155.4.457.

—— and W. A. Spencer. "Cellular Neurophysiological Approaches in the Study of Learning." *Physiological Review* 48, no. 1 (1968): 65–134. DOI: 10.1152/physrev.1968.48.1.65

Keleman, K., S. Krüttner, M. Alenius, and B. J. Dickson. "Function of the Drosophila CPEB Protein Orb2 in Long-Term Courtship Memory." *Nature Neuroscience* 10, no. 12 (2007): 1587–93.

Kellendonk, C., E. H. Simpson, H. J. Polan, G. Malleret, S. Vronskaya, V. Winiger, H. Moore, and E. R. Kandel. "Transient and Selective Overexpression of Dopamine D2 Receptors in the Striatum Causes Persistent Abnormalities in Prefrontal Cortex Functioning." *Neuron* 49 (2006): 603–15. DOI: 10.1016/j.neuron.2006.01.023.

Levine, A., Y. Huang, B. Drisaldi, E. A. Griffin, D. D. Pollak, S. Xu, D. Yin, C. Schaffran, D. B. Kandel, and E. R. Kandel. "Molecular Mechanism for a Gateway Drug: Epigenetic Changes Initiated by Nicotine Prime Gene Expression by Cocaine." *Science Translational Medicine* 3, no. 107 (2011): 107ra109. DOI: 10.1126/scitranslmed.3003062.

Majumdar, A., W. Colón-Cesario, E. White-Grindley, H. Jiang, F. Ren, M. R. Khan, L. Li, E. M. Choi, K. Kannan, F. Guo, J. Unruh, B. Slaughter, and K. Si. "Critical Role of Amyloid-Like Oligomers of Drosophila Orb2 in the Persistence of Memory." *Cell* 148, no. 3 (2012): 515–29.

Martin, K. C., A. Casadio, H. Zhu, E. Yaping, J. C. Rose, M. Chen, C. H. Bailey, and E. R. Kandel. "Synapse-specific, Long-Term Facilitation of Aplysia Sensory to Motor Synapses: A Function for Local Protein Synthesis in Memory Storage." *Cell* 91 (1997): 927–38. DOI: 10.1016/S0092-8674(00)80484-5.

Pavlopoulos, E., S. Jones, S. Kosmidis, M. Close, C. Kim, O. Kovaler-chik, S. A. Small, and E. R. Kandel. "Molecular Mechanism for Age-Related Memory Loss: The Histone-Binding Protein RbAp48." *Science Translational Medicine* 5, no. 200 (2013): 200ra115. DOI: 10.1126/scitranslmed.3006373.

Rajasethupathy, P., I. Antonov, R. Sheridan, S. Frey, C. Sander, T. Tuschl, and E. R. Kandel. "A Role for Neuronal piRNAs in the Epigenetic Control of Memory-Related Synaptic Plasticity." *Cell* 149, no. 3 (2012): 693–707.

Si, K., Y. B. Choi, E. White-Grindley, A. Majumdar, and E. R. Kandel. "Aplysia CPEB Can Form Prion-Like Multimers in Sensory Neurons That Contribute to Long-Term Facilitation." *Cell* 140 (2010): 421–35. DOI: 10.1016/j.cell.2010.01.008.

——, M. Giustetto, A. Etkin, R. Hsu, A. M. Janisiewicz, M. C. Min-iaci, J. H. Kim, H. Zhu, and E. R. Kandel. "A Neuronal Isoform of CPEB Regulates Local Protein Synthesis and Stabilizes Synapse-Specific Long-Term Facilitation in Aplysia." *Cell* 115 (2003a): 893–904. DOI: 10.1016/S0092-8674(03)01021-3.

——, S. Lindquist, and E. R. Kandel. "A Neuronal Isoform of the Aplysia CPEB Has Prion-Like Properties." *Cell* 115 (2003b): 879– 91. DOI: 10.1016/S0092-8674(03)01020-1.

INDEX

Abstract Expressionism, 63–65
addiction, 19–22
adolescents: and family poverty, 47–49; and gender identity, 49–52; and sports-related brain injuries, 44
The Age of Insight (Kandel), 57–63
alcohol, 21
Altmann, Maria, 62
Alzheimer's disease, 25, 26(fig.), 28, 41–42, 54
animal models, 13; deprived infant studies, 48; drug use studies, 20–21; memory studies, 14, 15, 17–18, 25–27; and schizophrenia, 23–24
anti-Semitism, 31–32, 72–74
Aplysia (sea snail), 14, 63; memory studies, 14, 15, 17–18
art: Abstract Expressionists, 63–65; Austrian Expressionists, 58–61; and beholder's share, 60, 62–63, 65–66; and biology of perception, emotion, and empathy,

59–61; and brain science, 57–69; and creativity, 61–62; as distillation of experience, 66; New York School, 63–64; and reductionism, 63–68
Austria, 71–74
"Austria and National Socialism" (symposium), 71–74
autism, 54–55
awards and honors, 34, 58, 62, 73, 74, 87–95
Axel, Richard, 6, 78, 78(photo), 79

Beck, Aaron, 10
bipolar disorder, 55
Bloch-Bauer, Adele, 62, 67
Bollinger, Lee, 78, 79
brain: and addiction, 19–22; and art, 57–69; and the biological basis for psychiatry, 7–11; bottom-up vs. top-down processing, 60–61; *Charlie Rose Brain Series* (PBS series), 35–53; and consciousness, 37–38; and

INDEX

brain (*continued*)
creativity, 57, 59, 61–62;
disorders of, 9–11, 40–42,
53–56 (*see also* mental illness;
specific disorders); and gender
identity, 49–52; and memory (*see*
memory); psychological and
social injuries, 44–49; relationship to the mind, 8–9, 33–34,
37; and social issues, 42–52; and
sports-related injuries, 43–44;
and visual perception, 59–66.
See also specific brain regions
Braun, Emily, 62
Burda, Frieder, 63
Burda Museum, 63
Burmeister, Gerd, 87

Charlie Rose Brain Series (PBS
series), 35–53
children: dangers of institutionalization and family poverty,
45–49; gender identity, 49–52
choice, 39–40. *See also* decision
making
cocaine addiction, 19–22
cognitive behavioral therapy, 10
Columbia University, 5–7, 81;
Howard Hughes Medical
Institute (HHMI), 6–7, 75, 81;
Jerome L. Greene Science Center,
76, 79–81; Kavli Institute for
Brain Science, 75–77; Mortimer B. Zuckerman Mind Brain
Behavior Institute, 76–79
concussion, 44
consciousness, 37–38
Costa, Rui, 78, 78(photo)
courts, 40
CPEB (cytoplasmic polyadenylation
element binding protein), 17–18

creativity, 57, 59, 61–62
*Cubism: The Leonard A. Lauder
Collection* (Metropolitan
Museum of Art publication),
62–63
"The Cubist Challenge to the
Beholder's Share" (Kandel essay),
62–63

D2 receptor, 24
decision making, 37, 38–40
dentate gyrus, 25–27, 26(fig.)
depression, 10, 48, 49, 55
The Disordered Mind (Kandel),
53–57
dopamine, 20–21, 23–24
Drosophila (fruit fly), 17
drug use, 13, 19–22

e-cigarettes, 22
economics, decision making in,
38–39
emotion, 39–40
entorhinal cortex, 25, 26(fig.)
epigenetics, 48. *See also* genes and
gene expression
European Neuroscience Society, 59
evolution, 34
Expressionism, 58–61, 63–65

Fischer, Heinz, 72–73
FosB gene, 21
framing, and decision making, 39
Fredrickson, Donald, 6
Freud, Sigmund, 58–61
fruit fly. See *Drosophila*

gambling experiment, 39
gateway effect in drug addiction,
19–22
gender identity, 49–52

genes and gene expression: and age-related memory loss, 25–26, 27(fig.); and cocaine addiction, 21; and learning, 9; and memory, 15; and mental illness, 9; and poverty, 48; and schizophrenia, 24
Gerstl, Richard, 67
Greene, Dawn M., 75–76
Grundfest, Harry, 5

hippocampus, 25, 26, 47–48
Holocaust, 72
hormones, and gender identity, 50–51
Howard Hughes Medical Institute (HHMI), 6–7, 75, 81
Huntington's disease, 41

In Search of Memory: The Emergence of a New Science of Mind (Kandel), 32–34
institutionalization, and brain injuries, 45–46

Jerome L. Greene Foundation, 75–76
Jerome L. Greene Science Center, 76, 79–81, 80(photo)
Jessell, Thomas, 6, 29, 78, 78(photo), 79
Jewish community in Vienna, 72
justice system, 40

Kahneman, Daniel, 39
Kandel, Denise, 1–2, 2(photo), 6, 22, 57, 59, 72–73, 84; and gateway effect in drug addiction, 19–22
Kandel, Eric, 2(photo); autobiographical sketch, 31–32; and *Charlie Rose Brain Series*, 35–53;

honors and awards, 34, 58, 62, 73, 74, 87–95; and Howard Hughes Medical Institute, 6–7, 81; interest in art and art collecting, 57, 59 (see also *The Age of Insight*; *Reductionism in Art and Brain Science*); and Jerome L. Greene Science Center, 79–81; and Kavli Institute, 77; and Mind Brain Behavior Institute, 78; move to Columbia University College of Physicians and Surgeons, 5–6; and Nobel Prize, 1–3, 30, 31–32, 83; in Paris, 74; and public understanding of science, 29–56; and symposium "Austria and National Socialism," 71–74; and symposium "The Long Shadow of Anti-Semitism at the University of Vienna Dating Back to 1870," 73–74; in Vienna, 72–73. *See also specific publications and research topics*
Kandinsky, Vassily, 64, 67–68
Kavli Institute for Brain Science, 75–77
Kellendonk, Christoph, 23
Klestil, Thomas, 71, 72
Klimt, Gustav, 58–59, 62, 67
Kokoschka, Oskar, 58–59, 67
Koslow, Stephen, 1
Kupfermann, Irving, 6

Lauder, Leonard A., 62
learning, 9, 20, 33, 55
Levine, Amir, 20
"The Long Shadow of Anti-Semitism at the University of Vienna Dating Back to 1870" (symposium), 73–74

Malevich, Kasimir, 64
Martin, Kelsey, 15
Mayberg, Helen, 10
memory, 13–19, 16(fig.), 20;
    ameliorating age-related memory
    loss, 25–28, 26(fig.), 27(fig.);
    defined/described, 33; disorders
    of, 13; initiation of, 14, 15;
    long-term memory storage,
    14–19; and poverty, 47; and
    schizophrenia, 24; short-term vs.
    long-term, 15, 16(fig.); and
    top-down processing, 61; and
    visual perception, 61, 65
mental illness, 9–11; biological
    markers for, 10–11; and *Charlie
    Rose Brain Series*, 40–42; as
    disturbance of brain function, 9,
    10 (*see also* brain: disorders of);
    and gene expression, 9, 21; and
    genetic, social, and developmen-
    tal factors, 9; mental health of
    transgender children, 49, 51–52;
    treatment, 9–10. *See also*
    addiction; depression; psychia-
    try; psychology; schizophrenia
Merton, Robert, 2–3
mice: deprived infant mouse
    studies, 48; drug use studies,
    20–21; memory studies, 17,
    25–27; schizophrenia studies,
    23–24
mind: consciousness, 37–38;
    principles of new science of the
    mind, 33–34; relationship to the
    brain, 8–9, 33–34, 37
Mind Brain Behavior Initiative,
    75–76, 78, 81
Mondrian, Piet, 64
moral philosophy, decision making
    in, 38–40

Mortimer B. Zuckerman Mind
    Brain Behavior Institute,
    76–79
mRNA, 15, 17, 18

Nazis, 72
Nelson, Charles, 46
neural circuits: and brain
    disorders, 54, 55; and changes
    in gene expression resulting
    from learning, 9; and depres-
    sion, 10, 55; and the mind,
    34; and visual perception,
    60–61
neurotransmitters, 14, 20–21,
    23–24
New York School, 63–64
nicotine, 19–22
Nobel Prize, 1–3, 30–32, 83
nucleus accumbens, 20

orphanages, 46

Papadopoulos, Elias, 25
parenting, 47–49
Paris, 74
Parkinson's disease, 41–42,
    54, 55
perception, and art, 59–66
Piano, Renzo, 79, 80
post-traumatic stress, 45
poverty, 45, 47–49
Practitioners Club (NYC), 59
prefrontal cortex, 24
Principles of Neural Science
    (Kandel, Schwartz, and Jessell),
    29–30
prions, 17–19, 18(fig.)
prosopagnosia, 42
protein misfolding, 42. *See also*
    prions

proteins: and brain disorders, 54; and memory, 13–19, 25–27
psychiatry: biological basis for, 7–11; and *Charlie Rose Brain Series*, 40–42; history of, 7–8
psychology: cognitive behavioral therapy, 10; and gender identity, 49; psychoanalysis, 7–8; psychological and social brain injuries, 44–49; psychotherapy, 9–11
public policy, 43

rationality, and decision making, 39
RbAp48 gene, 25–28, 27(fig.)
The Red Gaze (Schoenberg painting), 69(fig.)
reductionism, 63–68
Reductionism in Art and Brain Science (Kandel), 57, 63–68
reward system, 20–21
Richter, Gerhard, 63
right anterior insula, 10
Rokitansky, Carl von, 58–59
Romania, orphanages in, 46
Rose, Charlie, 35, 40. See also *Charlie Rose Brain Series*
Rothko, Mark, 63
Rowland, Lewis, 5

Schiele, Egon, 58–59, 67
schizophrenia, 13, 23–24, 42, 54–55
Schoenberg, Arnold, 67–68, 69(fig.)
Schoenberg, E. Randol, 62, 66–67
Schwartz, James, 5, 6, 29–30
scientific humanism, 56
scientific method, 32–33

sea snail. *See Aplysia*
Shohamy, Daphna, 63
Si, Kausik, 14, 15, 17
Siegelbaum, Steven, 6
Simpson, Eleanor, 23
Small, Scott, 25
smoking, and cocaine addiction, 19–22
social issues, and brain science, 42–52
Spencer, Alden, 5, 6, 29
Spitz, Rene, 45
sports, and brain injuries, 43–44
Stadler, Friedrich, 71, 72
Stern, Fritz, 71
stress, effect on the brain, 44–49
striatum, 20, 21, 24
Struhl, Gary, 6
suicidal thoughts, 48
synapses, 14–15, 54

transgender children, 49–52
transgenes, 24
trauma, 44–49, 55
Tversky, Amos, 39

University of Vienna: "Austria and National Socialism" symposium, 71–74; "The Long Shadow of Anti-Semitism at the University of Vienna Dating Back to 1870" symposium, 73–74; School of Medicine, 58, 74

vaping, 22
Varmus, Harold, 31
Vienna, 71–74. See also University of Vienna
Vienna School of Art History, 60
vision. *See* perception, and art

Woman in Gold (film), 62
Wyngaarden, James B., 6–7

Zeilinger, Anton, 71–73
Zuckerkandl, Berta, 58–59

Zuckerkandl, Emil, 58–59
Zuckerman, Harriet, 2–3
Zuckerman, Mortimer B., 76
Zuckerman Mind Brain Behavior
    Institute, 76–79